Animal Sonar Systems

NATO ADVANCED STUDY INSTITUTES SERIES

A series of edited volumes comprising multifaceted studies of contemporary scientific issues by some of the best scientific minds in the world, assembled in cooperation with NATO Scientific Affairs Division.

Series A: Life Sciences

Recent Volumes in this Series

The series is published by an international board of publishers in conjunction with NATO Scientific Affairs Division

A	Life Sciences	Plenum Publishing Corporation
B	Physics	London and New York
C	Mathematical and Physical Sciences	D. Reidel Publishing Company Dordrecht, Boston and London
D	Behavioral and Social Sciences	Sijthoff & Noordhoff International Publishers
E	Applied Sciences	Alphen aan den Rijn, The Netherlands, and Germantown U.S.A.

Animal Sonar Systems

Edited by
René-Guy Busnel
Ecole Pratique des Hautes Etudes
Jouy-en-Josas, France

and

James F. Fish
Sonatech, Inc.
Goleta, California

SPRINGER SCIENCE+BUSINESS MEDIA, LLC

Library of Congress Cataloging in Publication Data

International Interdisciplinary Symposium on Animal Sonar Systems, 2d, Jersey, 1979.
    Animal sonar systems.

    (NATO advanced study institutes series: Series A, Life sciences; v. 28)
    Symposium sponsored by the North Atlantic Treaty Organization and others.
    Bibliography: p.
    Includes indexes.
    1. Echolocation (Physiology) – Congresses. I. Busnel, René Guy. II. Fish, James F.
III. North Atlantic Treaty Organization. IV. Title. V. Series.
QP469.I57 1979   599'.01'88   79-23074
    ISBN 978-1-4684-7256-1   ISBN 978-1-4684-7254-7 (eBook)
    DOI 10.1007/978-1-4684-7254-7

This work relates to Department of the Navy Grant N00014-79-G-0006
issued by the Office of Naval Research. The United States Government
has a royalty-free license throughout the world in all copyrightable
material contained herein.

Proceedings of the Second International Interdisciplinary Symposium on
Animal Sonar Systems, held in Jersey, Channel Islands, April 1–8, 1979.

© 1980 Springer Science+Business Media New York
Originally published by Plenum Press, New York in 1980
Softcover reprint of the hardcover 1st edition 1980
A Division of Plenum Publishing Corporation
227 West 17th Street, New York, N.Y. 10011

ORGANIZING COMMITTEE
-=-=-=-=-=-=-=-=-=-=-=-=-=-=-=-=-=-=-

- R.G. Busnel
  Laboratoire d'Acoustique Animale
  E.P.H.E. - I.N.R.A. - C.N.R.Z.
  78350 Jouy-en-Josas, France

- J.F. Fish
  U.S. Navy
  Naval Ocean Systems Center
  Kailua, Hawaii 96734, U.S.A.

- G. Neuweiler
  Department of Zoology
  University of Frankfurt
  D-6000 Frankfurt, F.R.Germany

- J.A. Simmons
  Department of Psychology
  Washington University
  St. Louis, Mo. 63130, U.S.A.

- H.E. Von Gierke
  Aerospace Medical Research Lab.
  Wright Patterson Air Force Base
  Dayton, Ohio 45433, U.S.A.

A C K N O W L E D G M E N T S

-=-=-=-=-=-=-=-=-=-=-=-=-=-

This symposium was sponsored by various Organizations, National and International. The organizing Committee would like to thank them and their representatives :

- North Atlantic Treaty Organization, N.A.T.O.

- Advanced Study Institutes Programme

- Office of Naval Research of the U.S.A.

- United States Air Force

- Volkswagen-Stiftung from F.R. of Germany

- Ecole Pratique des Hautes-Etudes, Laboratoire d'Acoustique Animale, I.N.R.A. - C.N.R.Z. - 78350 JOUY-en-JOSAS, France

Preface

Thirteen years have gone by since the first international meeting on Animal Sonar Systems was held in Frascati, Italy, in 1966. Since that time, almost 900 papers have been published on its theme. The first symposium was vital as it was the starting point for new research lines whose goal was to design and develop technological systems with properties approaching optimal biological systems.

There have been highly significant developments since then in all domains related to biological sonar systems and in their applications to the engineering field. The time had therefore come for a multidisciplinary integration of the information gathered, not only on the evolution of systems used in animal echolocation, but on systems theory, behavior and neurobiology, signal-to-noise ratio, masking, signal processing, and measures observed in certain species against animal sonar systems.

Modern electronics technology and systems theory which have been developed only since 1974 now allow designing sophisticated sonar and radar systems applying principles derived from biological systems. At the time of the Frascati meeting, integrated circuits and technologies exploiting computer science were not well enough developed to yield advantages now possible through use of real-time analysis, leading to, among other things, a definition of target temporal characteristics, as biological sonar systems are able to do.

All of these new technical developments necessitate close cooperation between engineers and biologists within the framework of new experiments which have been designed, particularly in the past five years.

The scientists who have been working on these problems in various fields (electronics experts, signal processors, biologists, physiologists, psychologists) have produced new and original results, and this second symposium furnished the opportunity of cross-disciplinary contacts permitting an evaluation of the state of present research.

    The Jersey meeting in April, 1979, brought together more than
70 participants from 8 different countries. This meeting was particularly necessary considering the number of new research groups
that have appeared in various fields. I mention in particular:
the Federal Republic of Germany, where two young scientists who
participated at the Frascati meeting, Gerhard Neuweiler and Hans-
Ulrich Schnitzler, have since become professors and have founded
two schools of highly productive research; the United States, where
funded by the U.S. Navy, studies on dolphins have spread to San Diego
and Hawaii, given impetus by Bill Powell, Forrest Wood, Sam Ridgway,
C. Scott Johnson, Bill Evans, and Ron Schusterman, and where certain
of Donald Griffin's most gifted students, such as Alan Grinnell and
Jim Simmons, are continuing his work on bats at various universities;
Canada, where Brock Fenton is performing outstanding research.

    Although much research has been carried out in the Soviet Union
since 1969-1970, it is most unfortunate that, for reasons independent
of their wishes, our colleagues from this country, who have moreover published several excellent reviews of their work, were not able
to participate in our discussions.

    The tendancy which appeared at Frascati towards a certain zoo-
logical isolation corresponding to a form of animal specialization
dominated by either dolphins or bats, has partially regressed, thanks
to several physicists who use both bat and dolphin signals in their
theoretical approaches. While this segregation by field still remains a dominant behavior made obviously necessary, up to a certain
point, by the different biological natures of these two groups of
animals, the phenomenon is aggravated by the use of semantics specific to each group. Nevertheless, common interests and attempts at
mutual understanding which appeared are encouraging and should be
congratulated.

    The proceedings of the Jersey meeting demonstrate the extent
to which technology has advanced in the past decade, in performances
of transducers, various microphones, hydrophones, as well as in
recording apparata, analytical methods, particularly in the use of
signal processing techniques, and in the application of new ideas
such as time-domain, auditory processing, frequency-domain, Doppler
compensation, target-acoustic imaging, and so on.

    It is also interesting to note to what extent experimental
strategies have been refined, and one can only admire the elegance
of certain demonstrations carried out on dolphins as well as on
bats. Many aspects of the performances of diverse species continue,
however, to intrigue us, as they reveal sensory abilities whose
fine analysis still eludes us, particularly the central mechanisms
which control and regulate them. As an example of this, I would
particularly like to mention how enriching was the experience that
we were able to have using Leslie Kay's apparatus for the blind,

which gave spatiodirectional sensations analogous to those of air-
borne animal sonar systems. I do not doubt in the least that those
several minutes will lead to a totally new concept of biosonar problems.

As Henning von Gierke pointed out during the last session and
in a personal communication, the trends appearing at the Jersey
meeting indicate that pattern recognition theory is becoming more
and more important to biosonar research, replacing range finding
and echo theories as the promising research areas of the future.
Pattern recognition specialists should be included in future meet-
ings as well as experts on the spatial frequency analysis of acoustic
and visual perception. Although animal sonar might be used to a
large extent for acoustic imaging of space, we know from ultrasono-
graphy that acoustic images are different from optical images.
Acoustic space perception therefore differs from visual space perception. Since the acoustic space is scanned sequentially, the
total "acoustic image" depends primarily on memory capability, which
is the major difficulty encountered by Leslie Kay in his device for
the blind. The findings of this Symposium may have a major impact
on general auditory physiology regarding, on one hand, the example
of sharp filter (acoustic fovea) and on the other, peripheral processing.

A reading of the present set of volumes presenting the current
state of research will bring out at the same time the unknowns of
the problem, the uncertainties, the hypotheses, and will allow vet-
erans of Frascati to measure the progress made since then.

I am most happy to thank here my American colleagues who agreed
to respond to my call in 1977, particularly Henning von Gierke, who
was our referee and support for N.A.T.O. as well as the U.S. Air
Force, and Bill Powell and Forrest Wood for the U.S. Navy, who later
recommended to me Jim Fish, a young associate, dynamic and efficient,
who held a preponderant place in the Organizing Committee.

I also would particularly like to express my gratitude to Gerhard
Neuweiler for his constructive participation in our planning group.
It is thanks to his outstanding reputation that our Symposium was
able to obtain funding from Volkswagen for the active and highly
productive German delegation. He assumed as well the heavy respon-
sibility of financial management of our funds.

I also wish to thank, on behalf of our Committee and myself,
the diverse individuals from my laboratory who, with devotedness,
took on countless tasks, often thankless and lowly: Michèle Bihouée,
Annick Brézault, Marie-Claire Busnel, Sophie Duclos, Diana Reiss,
and Sylvie Venla. Leslie Wheeler, who assisted me in editing these
volumes, deserves special mention, as without her help I would not
have been able to publish them so rapidly.

During the last plenary session in Jersey, the Organizing
Committee and the Co-chairmen of the different sessions decided on
the publication format of the proceedings, and suggested to the
Symposium participants to dedicate this book to Donald R. Griffin.
Our colleagues unanimously rendered homage to the spiritual father,
the inventor, of echolocation.

The scientific wealth brought out during the three half-day
poster sessions, bears witness to the interest and importance of
the work of numerous young scientists and makes me optimistic for
the future of the field of animal sonar systems. For this reason,
and thanks to the two experiences which the majority of you have
considered successful, I wish good luck to the future organizer of
the Symposium of the next millenium.

René-Guy Busnel
Jouy-en-Josas
France

Contents

On behalf of the members of the organizing Committee and all the Symposium participants, we dedicate this book to

Donald R. GRIFFIN

in homage to his pioneering work in the field of echolocation.

Donald R. Griffin

DEDICATION

by Alan D. Grinnell

Just over 40 years ago, a Harvard undergraduate persuaded a
physics professor to train his crystal receiver and parabolic horn
at active bats. They detected ultrasonic pulses, and the contemporary field of echolocation research was born. The undergraduate was
Donald R. Griffin, the professor, G. W. Pierce, and their first report appeared in 1938. In recognition of his founding role and his
countless important contributions to echolocation research this volume is dedicated to Don Griffin. Fortuitously, this coincides with
the approach of his 65th birthday, a time when particularly popular
and influential figures in a field are often honored with a "Fest-
schrift" volume.

We were fortunate, at this meeting, that Don Griffin was persuaded to add a few recollections of his experiences during the in-
fancy of the field. These are included in this volume. Additional
perspective on some of the early years has been volunteered by his
partner in the first experiments demonstrating echolocation, Dr.
Robert Galambos, then a Harvard graduate student, now a well-known
auditory neurophysiologist and Professor of Neuroscience at UCSD.

"In early 1939 Don found out that Hallowell Davis at the
Harvard Medical School was teaching me how to record electric
responses from guinea pig cochleas and asked if I could slip
in a few bats on the side. He wanted to know whether their
ears responded to the "supersonic notes" he and Professor G. W.
Pierce had just discovered. So I asked Dr. Davis, who said
"go ahead" and thus Don found himself a collaborator in some
unforgettable adventures. (This was not the only time Don
enlisted me in one of his enterprises--I remember spending
3 days helping him build a bird blind on Penekeese Island

in weather so foul the Coast Guard had to come rescue us.)

During that spring I worked out the high frequency respon-
sivity of the bat cochlea, and in the fall we assessed the
obstacle avoidance capabilities of the 4 common New England
species (using an array of wires hanging from the ceiling to
divide a laboratory room into halves). We recorded the inaudible cries bats make in flight and demonstrated (by inserting
earplugs or tying the mouth shut) that they must both produce
and perceive them if their obstacle avoidance is to be successful. (We also made a sound movie of all this which nobody can
find.) That research yielded my Ph.D. thesis and launched Don
on the career this book honors.

All the crucial new measurements we made used the unique
instruments devised by Professor Pierce. A physics professor
who, like Don, seemed to me really a naturalist at heart, Pierce
had designed his "supersonic" receivers in order to listen to
the insects singing around his summer place in Vermont (or was
it New Hampshire?). How Don found Professor Pierce and then
talked him into letting us use his apparatus I do not know.

In the spring of 1940, when we felt we understood how bats
avoided obstacles in the laboratory, it occurred to us to test
our ideas in the field. So we made an expedition to a cave
Don knew in New York State taking along the portable version of
Pierce's supersonic receiver shown in the accompanying picture
(Don was a skillful photographer even in those days).

The cave in question opened just beyond the bank of a
mountain brook. Once inside you first climbed down a ways and
then up. We had no trouble getting in and, after the small
upward climb, found ourselves looking along a straight tunnel
as far as our flashlights could penetrate. We set up the
equipment and Don went further on to the gallery where the bats
roosted. He sent several of them flying, one by one, down the
tunnel in my direction and as they approached Professor Pierce's
machine emitted the chattering clamor we hoped for. Then Don
and I changed places so he could hear the noise too. Since
the readout of Pierce's portable device came only via earphones,
the sole record of the first bat ultrasonic cries ever heard
outside a laboratory is the one engraved in our memories.

Throughout all of this Don kept urging me not to waste
time, and I thought he closed off the exercise and moved us
toward the entrance after an almost indecently short interval.
He explained his behavior once we had scrambled down and then
up and then out to cross the brook: the spring sun was rapidly
melting the snow, the brook was rising fast, and he had been
worried from the moment we arrived that the water might fill

the entrance to the cave and trap us inside. My captain, I
thought to myself, always looks out for the safety of his crew."

.During the nearly four decades since that first demonstration
of echolocation, the field has flourished. This is evident from
the quality and diversity of work described in this volume. It can
also be seen, quantitatively, in the accompanying graph, which shows
the number of publications on echolocation in microchiropteran bats
during each 3-year period since 1938. This graph, kindly prepared
by Uli Schnitzler, shows several surges in productivity - one about
1961, when the second generation of U.S. scientists began publishing,
and a very large one in 1967, when, following Frascati, the second
generation of German scientists and the Russian group entered the
field. Now, with the Jersey Symposium, the third generation is beginning to contribute importantly. Throughout this entire period,
while moving professionally from Harvard to Cornell, back to Harvard,
and then to the Rockefeller University, and while making major contributions to a number of other fields, Don Griffin has continued to
guide development of the field of echolocation with imaginative, incisive experiments: demonstrating the usefulness of echoes for orientation in the lab and in the field, showing that echolocation was
used for insect capture, documenting the sensitivity of the system
and its resistance to jamming, developing techniques to show how
accurately targets can be discriminated, and analyzing the laryngeal

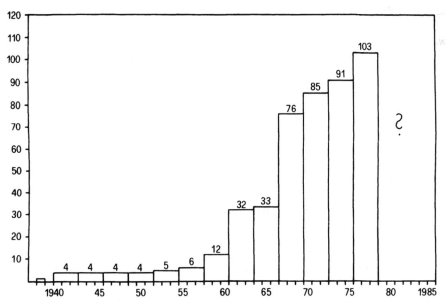

Histogram of papers published per 3-year period since 1938 on echolocation in microchiropteran bats.

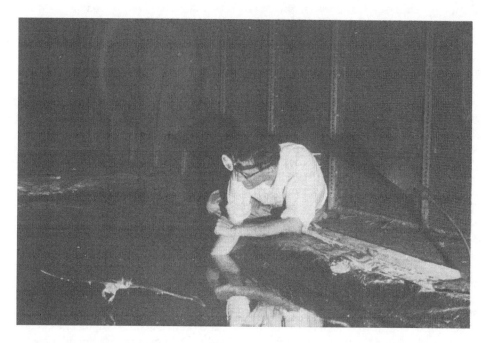

D. R. Griffin at Simla, Trinidad, studying capture of food by
<u>Noctilio</u> (1960).

mechanisms involved in ultrasonic orientation. It was in his lab,
with his encouragement, that the first major advances were made in
understanding the neural adaptations for echolocation, an aspect of
the field that has grown to enormous and impressive proportions. In
association with Al Novick, he established the great value of comparative approaches to echolocation, traveling the world not only to
document the variety of sounds and skills shown by bats, but to do
the first careful studies of echolocating birds, as well. Out of
early interests in the feeding habits, home ranges, and seasonal mi-
grations of bats (not to mention a long-standing fascination with
homing and migration in birds) came pilot studies on the role of
echolocation, passive hearing and vision in homing behavior. Indeed,
in countless instances, he has broken new ground and set standards
of rigor and experimental elegance that have served the field well.

A major milestone in the field of animal behavior was the publication, in 1958, of <u>Listening in the Dark</u>, Griffin's monograph on
his experiments and thoughts on echolocation, and winner of The Daniel
Giraud Elliot Medal of the National Academy of Sciences. With the
perspective of 20 years subsequent work in the field, it is astonishing to anyone re-reading this classic to realize how fully Griffin
already understood the phenomenon of echolocation, how many critical
discoveries he had already made, and how profound were his insights.
This was one of five books Don Griffin has written, the most recent

of which is "The Question of Animal Awareness: Evolutionary Continuity of Mental Experience" (1976).

It was not only for his incisive early work that Don Griffin deserves recognition as "father" of the field (or Godfather as he was also described during the meeting); he has also been the academic father, or colleague, of a high percentage of those who have become active in the field. Most of the contributors to this volume who have worked on bats have felt the imprint of Griffin's personality and experimental approach directly, as graduate students or post-doctoral research associates working in his laboratory, or as their students, and it has been a powerful influence on their lives and careers.

Don is a great storyteller, as his chapter in this volume attests. More than that, however, his approach to science is guaranteed to never leave a dull moment. Whether it is fighting to stay aloft after releasing birds from a small plane piloted by Alex Forbes, or trying to convince suspicious authorities that he had valid reason for clamboring around the fire escapes of a mental hospital, or barely escaping a flooded cave, or covering for a student who had fallen through the ceiling of the Mashpee Church, or searching the tombs of an ancient Italian cemetery for Rhinolophus, his approach to experimentation is as direct and audacious as his ideas. Many of the breakthroughs in echolocation research were the result of his introduction of new or unfamiliar technology, from home-made ultrasonic microphones to high speed tape recorders, sonagraphs, and information theory. This applies in all of his other fields, as well. In recent years, for example, much of his energy has been devoted to radar identification and tracking of migrating birds. One of his collaborators in this research, Charles Walcott, tells of just one of the complications that this has gotten them into, and of Don's characteristic aplomb in solving the problem:

"That season Don arrived in Stony Brook with a trailer carrying what looked like a giant coffin. Actually the box contained a helium filled kitoon - a sort of hybrid balloon - kite combination that could be flown up to several thousand feet in the air. Its purpose was to place instruments in the same air mass that the radar showed birds to be flying in. Indeed on its trials in Stony Brook the kitoon worked splendidly - we were able to get it up at least a thousand feet or more - amply high enough to be where birds flew. Unfortunately they aren't the only things that fly there - a helicopter soon appeared flying significantly lower than the kitoon. As it came close and inspected the kitoon we noticed that the helicopter bore the inscription "Police" in large letters. We rapidly hauled down the kitoon and anxiously awaited the arrival of sirens and patrol cars; fortunately they never came. But on other occasions, the law has become

involved and Don had been read the Federal Air Regulations
which prohibit the flying of balloons, kites or other such
devices more than a few feet above ground. The State Police
in Millbrook, New York next to The Rockefeller field station
had alerted the FAA about Don's activity. Don promised the
FAA representative that it wouldn't happen again. It was the
following Easter Sunday when Don and his colleagues were
following a kitoon with radar that the kitoon's string broke.
Following the kitoons progress with the radar, they saw that
it suddenly stopped drifting in the wind and became stationary.
It was hanging only a few hundred feet in the air directly over
the State Police headquarters. The string had caught in the
upper branches of a small sapling. With a typical display of
ingenuity, Don rigged a second kitoon with a grappling iron
and managed to snag the string of the escaped kitoon retrieving
both just before dawn on Easter Sunday!"

To those of us who have had the privilege of working with Don
Griffin, perhaps the greatest lesson has been his emphasis on rigor
in experimentation. No one is more acutely aware of the ambiguities
of an experiment, the difficulties involved in demonstrating some-
thing convincingly. Nor does he ever jump to conclusions. His intellectual vigilance has been characterized by the claim that if he
were in a car passing a flock of sheep in a field, and a travelling
companion commented on the fact that among the sheep were two that
were black, he would reply, "They're black on the side facing us,
anyway." (Attributed to Don Kennedy.) On the other hand, he is not
afraid to consider some of the most complex formsof animal behavior,
and to take on established dogma, decrying the excessive use of "simplicity filters" in interpreting behavior, a position he argues elo-
quently in his recent writing on animal awareness.

This mixture of insistence on careful unbiased observation of
what is really there and rigorous proof of any conclusions, combined
with a brilliant imagination and willingness to adapt new technology
to biological problems, has done much to develop the field of animal
behavior research. It has served as particular inspiration for all
of us.

We are honored to dedicate this volume to Donald R. Griffin.

MODELS FOR ECHOLOCATION

Richard A. Altes

ORINCON Corporation

3366 N. Torrey Pines Ct., La Jolla, CA 92037, USA

I. INTRODUCTION

Mathematical or engineering models of biological systems are viewed with extreme skepticism (if not with hostility) by some experimental biologists. A discussion about models should take account of this attitude, since the work may otherwise be doomed to obscurity, even if the results are correct. The skepticism about modelling can be attributed to one or more of the following arguments:

1. Models are unnecessary if experiments result in a good intuitive and verbal understanding of the functions that are performed by a biological system. If animal behavior and physiology can be understood in non-mathematical terms, the value of a mathematical model becomes questionable. "Physics is mathematical not because we know so much about the physical world, but because we know so little; it is only its mathematical properties that we can discover" (B. Russell, 1927). In this respect, biology may be fundamentally different from physics.

2. Models that are based upon insufficient data can be misleading, and it is not clear when the data base is sufficiently large to justify the formulation of a model. In A. C. Doyle's "A Scandal in Bohemia," Sherlock Holmes tells Dr. Watson: "It is a capital mistake to theorise before one has data. Insensibly one begins to twist facts to suit theories, instead of theories to suit facts." Even in engineering and physics, theories often rest upon assumptions that are "tractable" but overly simplistic.

3. Models should improve our comprehension of nature by using engineering concepts to interpret biological data, but the end result may be an overly restrictive or narrow viewpoint. Like philosophies and religions, models suggest a definite conceptual framework for the interpretation of nature, and the adoption of one such structure can preclude other interpretations. Engineers and physicists also have a tendency to favor theories that possess beauty, order, or simplicity. This attitude is nicely expressed by the physicist P. Dirac (1963). A predisposition toward mathematical elegance can again   result in a viewpoint that is too restrictive for biology.

4. Aside from possible improvement of man-made systems, there is sometimes no obvious benefit to be gained by constructing a model. Model makers often fail to suggest experiments to test their constructions, and these experiments are of primary importance to biologists. The result is an elaborate theory that is based largely upon conjecture--a house of cards without a foundation of check points or experiments that can be used to verify its relevance to nature.

The above list of objections to mathematical formulations can be used by the model maker to identify weaknesses in his exposition. Indeed, it will become apparent that items from the list apply to some past and present models for animal echolocation. In spite of the above objections, the following arguments seem to justify a modelling approach.

1. Models are a means for cross-fertilization between biology and engineering. The importance of this effect is described by A. Koestler in The Act of Creation (1964): "All decisive advances in the history of scientific thought can be described in terms of mental cross-fertilization between different disciplines." Examples in the study of animal echolocation:

(i) The mystification of L. Spallanzani concerning the interpretation of his bat navigation experiments, due to the fact that the wave theory of sound (and thus the concept of ultrasound) had not been discovered (Griffin, 1958).

(ii) Radar theory indicates that the E-E neurons of Menabe, Suga, and Ostwald (1978) can be used not only for detection, but also for estimation of echo strength (as discussed in Part VI of this section). This idea implies that detection and estimation may be performed by the same neuronal components.

2. Models provide a mechanism for expanding our conceptual framework and our vocabulary for describing biological data. For example:

(i) A study of angle estimation from the viewpoint of binaural processing (Altes, 1978a) indicates that sound localization depends not only upon beamwidth of transmitter and receiver (i.e., upon the amplitude of an array pattern) but also upon the phase of this pattern as a function of frequency and angle. (We are monaurally insensitive to a constant phase shift, but not to a phase shift that varies with frequency.) This concept suggests that many physical measurements of transmitted beam width and directional response of the pinna are incomplete.

(ii) The discovery of excitatory-center, inhibitory-surround receptive fields for sound localization in the midbrain of the owl (Knudsen and Konishi, 1978) can be related to vision and Von Bekesy's "neural sharpening" ideas. But such receptive fields can also be interpreted as part of a principal component analysis for spatial pattern recognition (to be discussed in this section), or as part of a sequence of spatial filters that are sensitive to the acoustic counterpart of lines, edges, and more gradual contrast changes (Altes, 1976a; Hauske, Wolf, and Lupp, 1976).

3. A model provides a method for evaluating the significance of specific biological data. For example:

(i) Suppose that we have a set of neurons that fire in response to a received echolocation pulse, but the firing times exhibit a great deal of jitter. A model which forms a best estimate of pulse arrival time reveals that a very accurate arrival time estimate can be obtained by averaging statistically independent neuronal responses. The variance of the arrival time estimate is inversely proportional to the number of independent neuronal responses that are averaged together.

(ii) Green (1958) and Green, McKey, and Licklider (1959) have deduced a summation law for auditory detection of multiple sinusoids in humans. Detectability depends upon the sum of the squares of the energies of the different components. This behavior is found in a spectrogram correlation process, which is a locally optimum detector for signals with known covariance function (Altes, 1978b). This detector and its relevance to echolocation will be discussed further on.

(iii) An empirically determined equation for just-noticeable frequency difference in humans (Siebert, 1970) is commensurate with the use of a spectrogram correlator for frequency discrimination.

4. A definitive model can transform a conjecture into a hypothesis that can be experimentally tested, by predicting a particular
aspect of animal behavior. If the prediction is supported by data,
the conjecture may be correct. Models thus provide the experimenter
with a feedback mechanism, such that conjectures from past data can
be transformed into meaningful future experiments. Examples:

(i) An energy spectrum analyzer (Johnson and Titlebaum,
1976) is particularly susceptible to noise with a rippled
power spectrum, e.g., noise added to a delayed version of
itself (Johnson, 1972). A broadband matched filter is relatively unaffected by this kind of interference. This difference can be used as the basis for jamming experiments with
bats and cetaceans, to differentiate between models.

(ii) If a spectrogram correlator is used for estimation of
time delay, then we expect a just-noticeable delay difference
$\Delta\tau$ that varies as $B^{-\frac{1}{2}}$, where B is the bandwidth of the echolocation signal (Altes, 1978b). For estimation with a matched
filter, $\Delta\tau \sim B^{-1}$. Once again, one can construct experiments to
test these hypotheses (spectrogram correlator versus matched
filter).

5. A model provides a test of one's understanding if it can
actually be built, or if its performance can be simulated on a computer. It can then be seen whether the model really "works" for realistic data, a necessary condition for the validity of the conceptual
framework. For example, the geometrical theory of diffraction (Bechtel,
1976) along with concepts of Doppler tolerance (Altes and Reese,
1975) or simplicity of implementation (Altes, 1976b) suggest that
target echoes should be divided into highlights, and that each highlight should be characterized by an echo $f_n(j\omega)^{n/2}U(\omega)$, where f_n is
a constant, $U(\omega)$ is the transmitted signal in the frequency domain,
and n is an integer between -2 and +2. A software implementation
with real echo data, however, has revealed important deficiencies
in this approach (Skinner, Altes, and Jones, 1977) and has necessi-
tated a search for a different model of target discrimination capability in animal echolocation.

6. Although the concept of elegance or beauty has been listed
as a disadvantage in applying mathematical models to biology, the
quest for elegance and a unifying conceptual framework must also be
listed as a fundamental motivation for much scientific inquiry
(Koestler, 1964). An "inner beauty" is often found in a mathematical
formulation that is truly descriptive of reality, "but sometimes
the truth is discovered first and the beauty or 'necessity' of that
truth seen only later" (Feinman, 1974). Mathematical models provide
a pathway from a confusing melange of data to a "beautiful"--and
hopefully simple--conceptual framework.

The remainder of this section describes some mathematical
models that have recently been applied to animal echolocation.

Part II describes models that are associated with correlation or
matched filtering (followed by envelope detection) and the point
target assumption. Part III discusses the use of correlation processing for range-extended targets. Part IV involves a description
of a target in terms of a scattering function or a time-frequency
energy density function, and the use of spectrogram correlation to
detect and classify such targets. Part V describes estimation and
detection performance when spectrograms are used to describe an
echo, and behavioral tests that can be used to distinguish between
spectrogram correlation and matched filtering. Part VI discusses
theories of binaural interaction and sound localization, and Part
VII is concerned with the issue of phase sensitivity in echolocation.
Part VIII discusses the results in terms of the objections to modelling that were expressed at the beginning of this introduction.

II. CORRELATION OR MATCHED FILTERING FOLLOWED
     BY ENVELOPE DETECTION

a. Elementary Target Models

     The simplest assumption concerning a radar/sonar reflector is
that it behaves as a point target. The echo from a motionless point
target is an exact replica of a signal that is transmitted from a
motionless sonar, delayed in time. If the target is modelled as
a linear filter, then the impulse response of this filter is
assumed to be an impulse. The point target assumption can be
applied to range-extended targets by assuming that the target echo
is composed of a sequence of highlights or specular reflections
(Freedman, 1962), and that each of these highlights behaves as a
point target.

     A point target model implies that the echo from a target with
zero range rate is exactly known except for delay, if the transmitted signal is known. In a background of white, Gaussian noise,
the ideal detector for a point target echo is then a matched filter
or correlator (Woodward, 1964; Cook and Bernfeld, 1967).

     A slightly more general point target model assumes that the
target impulse response is a delta function multiplied by a complex constant with unknown phase. The echo is then known exactly
except for delay and phase. In this case, a non-adaptive, locally
optimum echo detector consists of a matched filter or correlator
followed by an envelope detector (Altes, 1979b). A locally optimum
detector is one that is optimized for low signal-to-noise ratio
(SNR), where optimality is most important (Capon, 1961; Middleton,
1966). Such a detector may be non-ideal for higher SNR, but sub-
optimal behavior in such cases is often permissible. If a sequence
of echoes is available, and if each echo has the same unknown phase,

one can sequentially estimate the probability density function (pdf) that describes the phase, on the basis of accumulated echo data. This updated pdf can be used to adaptively improve the performance of the detector for each new echo (Roberts, 1965; Hodgkiss, 1978).

Matched filtering is also relevant to any extended target that can be modelled as a time-invariant linear filter. In this case, the echo is determined by convolving the transmitted signal with the target impulse response. A reasonable way to measure the impulse response is to transmit an approximation to an impulse. Because of transmitter power limitations, however, it is often preferable to transmit a broadband chirp. The response of a filter that is matched to the transmitted signal then approximates the target impulse response, if the signal is sufficiently broadband (Turin, 1957; Altes, 1977).

b. Doppler and Acceleration Tolerance

The maximum matched filter response to a point target or to a sequence of highlights is reduced when the signal is scaled because of target motion. A filter that is matched to the original signal is mismatched to the scaled signal. One can exploit this effect to both detect the echo and to measure target velocity by using a bank of filters, each of which is matched to a differently scaled version of the transmitted signal (Cook and Bernfeld, 1967; Rihaczek, 1969; Vakman, 1968). To detect a Doppler-scaled echo without resorting to a whole bank of filters, a Doppler-tolerant signal can be used. When a Doppler-tolerant signal is transmitted, the mismatch between the scaled signal and the original signal does not cause significant degradation of the envelope detected matched filter response.

A graphical derivation of such a signal (Altes and Titlebaum, 1975) is shown in Fig. 1. A picture of the transmitted signal is projected from P onto a screen at OT. A time-compressed picture of the signal (corresponding to an echo from a target that is getting closer to the sonar) is obtained by inserting a second screen closer to the projector. The last zero crossing of the unscaled signal occurs at time T on the far screen, and this zero appears at time T/s on the near screen. For Doppler tolerance, the zero at T/s should have a counterpart at time A = T/s on the far screen. The zero at time A projects back and intersects the near screen at time B. A zero should therefore be found at time C = B on the far screen. This process shows that the distance $T_i(t)/2$ between a zero crossing at time t and the next zero crossing is

$$T_i(t)/2 = (s-1)t \qquad\qquad\qquad (1)$$

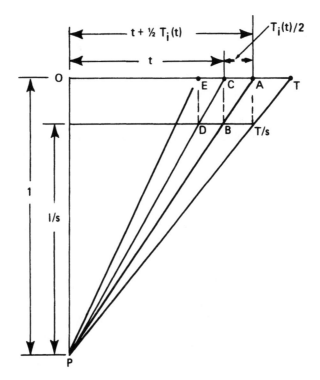

Fig. 1. Graphical derivation of zero crossings
for Doppler tolerance.

where s is the scale factor that is introduced by target motion,
i.e.,

$$s = (1+v/c) \ / \ (1-v/c) \approx 1 + 2v/c \ . \tag{2}$$

In (2), v is range rate and c is the speed of sound.

Suppose that a signal with zero crossings as in (1) is scaled
by a factor s_1 rather than s, where $s_1 = 1 + v/c$. The zero crossings of the scaled signal fall between those of the unscaled signal,
i.e., a third screen is inserted midway between the two screens in
Fig. 1. If the unscaled signal is a frequency modulated cosine
function, then the signal with scale factor s_1 can be written as a
frequency modulated sine function. A signal with an arbitrary scale
factor can be written as a linear combination of a cosine function
and a sine function. This observation implies that, for Doppler
tolerance, the receiver should use a signal with zero crossings
described by (1) together with two correlators, one of which corresponds to a frequency-modulated cosine function and the other to
a frequency modulated sine function. The sum of the squares of the
correlator outputs is relatively insensitive to Doppler-induced
scale changes. The response of each correlator, however, is

Doppler sensitive, and target velocity can be estimated by computing the arctangent of [the sine correlator response divided by the
cosine correlator response], i.e., by measuring the _phase_ of the
echo (Altes and Skinner, 1977). This velocity estimate requires
only two matched filters, rather than the whole bank of filters
that was mentioned earlier.

In practice, a Doppler-tolerant receiver can be obtained by
passing the squared response of a single matched filter or correlator through a low pass filter or integrator. This envelope
detection process is approximately equivalent to summing the squares
of the outputs from the sine and cosine filters (Davenport and
Root, 1958).

A signal with zero crossings as in (1) has a period $T_i(t)$ that
is proportional to time, and it is called a linear-period modulated
(LPM) signal (Rihaczek, 1969; Kroszczynski, 1969). The instantaneous frequency is inversely proportional to $T_i(t)$, and it appears as
a hyperbola on a spectrogram plot. Such signals are employed by
many of the so-called FM bats (Cahlander, 1966; Griffin, 1958), and
they are used by other bats such as _Rhinolophus_ when these bats
are close to the target of interest. One can also model some
cetacean echolocation clicks with LPM signals, although the clicks
have relatively few zero crossings (Altes and Reese, 1975; Altes,
1976b).

The technique in Fig. 1 can also be used to obtain signals that
are both Doppler and acceleration tolerant. By tilting the near
screen as in Fig. 2 and following the same procedure as before, we
obtain the counterpart of (1) for targets with variable velocity,
viz.,

$$T_i(t)/2 = [s(t)-1]t .\tag{3}$$

When the change in velocity over the duration of the pulse is much
less than the initial velocity, we obtain (Altes and Titlebaum,
1975)

$$T_i(t) = t/(C_0+C_1t)\tag{4}$$

where C_0 and C_1 are constants. Eq. (4) describes an instantaneous
period that increases linearly with time and then levels off to a
constant as C_1t becomes much greater than C_0. The instantaneous
frequency first decreases in proportion to $1/t$ and then levels off
to a constant frequency toward the end of the pulse. Signals of
this type have been measured from _Lasiurus borealis_ in the laboratory (Cahlander, 1966) and from _Pipistrellus subflavus_ in its
natural environment (data obtained by Simmons and Griffin).

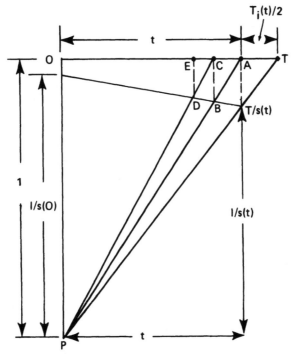

Fig. 2. Graphical derivation of zero crossings for
Doppler and acceleration tolerance.

c. Doppler Sensitivity

The above observations demonstrate a lack of uniqueness that
sometimes emerges from a theoretical interpretation of data. Having
found LPM signals in nature, we have been inclined to label the corresponding animal sonar system as Doppler-tolerant, although velocity can still be measured with more than one pulse. If a phase
measurement can be obtained, however, LPM signals can be used to
estimate radial velocity on the basis of a single echo.

Such a uniqueness problem can sometimes be solved by considering additional data and by careful analysis of theoretical assumptions. For example, a joint velocity and range measurement with
LPM signals is severely degraded unless range-Doppler coupling
errors are eliminated. The formulation of Altes and Skinner
requires LPM signals to be symmetric in time in order to eliminate
range-Doppler coupling, and bat signals do not display such symmetry. The original symmetry requirement, however, can be generalized,
and it can be shown that range-Doppler coupling is eliminated if an
LPM signal is preceded or followed by a constant frequency tone
with appropriate amplitude and duration. This situation occurs for
CW-FM bats, and for some "FM" bats such as Pipistrellus and

Lasiurus, which might better be labelled "FM-CW" bats. Signals that are Doppler and acceleration tolerant when a matched filter is followed by an envelope detector can thus also be Doppler sensitive, with no range-Doppler coupling, if phase is measured.

For frequencies above 6 kHz, the auditory nerve cannot encode the timing information that allows one to distinguish between a sine and cosine function (Anderson, Rose, Hind, and Brugge, 1971). It can therefore be argued that phase information is lost for ultra-sonic signals, and a velocity measurement is impossible. Neverthe-less, behavioral and neurophysiological evidence indicates that humans and cats are sensitive to the relative phase between two signals (Patterson and Green, 1970; Kotelenko and Radionova, 1975). The energy spectrum of a signal with Fourier transform $U(f)\exp(j\phi)$ is $|U(f)\exp(j\phi)|^2 = |U(f)|^2$, and the phase ϕ is eliminated. For the sum of two signals, however,

$$|U_1(f)\exp(j\phi_1) + U_2(f)\exp(j\phi_2)|^2$$

$$= |U_1(f)|^2 + |U_2(f)|^2 + 2\,\mathrm{Re}\{U_1(f)\,U_2^*(f)\exp[j(\phi_1-\phi_2)]\},$$

$$(5)$$

and the energy spectrum depends upon the difference $\phi_1-\phi_2$. Since bat signals often include harmonics, the relative phase shift between harmonics can be measured, and this quantity contains velocity information. Two more methods of phase measurement that do not require harmonics will be discussed in Part VII.

On the basis of past experimental data, it is not clear whether velocity can be measured from a single echo by bats that use LPM signals. This situation provides an example of the feedback from models to future experiments that was listed in the introduction. An appropriate behavioral measurement should be able to answer the questions: (i) Are FM bats sensitive to the difference between unscaled and scaled LPM signals, for very small scale changes? (ii) Does this sensitivity (if it exists) depend upon the presence of harmonics?

The possibility of phase measurement has been considered in the context of hearing models, but what about the possibility of matched filtering? Given that the peripheral auditory system behaves as a bank of overlapping filters that are followed by enve-lope detectors (Siebert, 1968; Evans, 1977), how can one justify a matched filter or correlator model? A matched filter followed by an envelope detector is, in fact, theoretically feasible because the original input signal can be reconstructed from its spectrogram, aside from a constant, unknown phase shift (Altes, 1978b,d).

d. Signal-Filter Design for Maximization of
    Signal-to-Interference Ratio

It would seem that any mathematical model for a biological
system should be able to adapt to environmental changes. In
sonar, the environment is characterized by noise and reverberation
(clutter) with particular statistical characteristics. Detector
performance in a given clutter environment can be measured in
terms of the signal-to-interference ratio (SIR), where

$$\text{SIR} = \cfrac{\text{Maximum detector output when input = target echo}}{\left\{ \begin{array}{l} \text{[Expected detector output when input = noise} \\ \text{(no signal transmitted)] + [Expected detector} \\ \text{output when input = clutter reflections} \\ \text{(reverberation)]} \end{array} \right\}} \qquad (6)$$

When the detector consists of a matched (or mismatched) filter
followed by an envelope detector, and when a point target model is
used, it is possible to find the best signal-filter pair for maxi-
mizing SIR. The resulting signal and filter functions depend upon
environmental clutter and noise conditions. The original work in this
area was performed in a radar context using narrowband signals (Delong
and Hofstetter, 1969; Stutt and Spafford, 1968; Rummler, 1966). The
radar formulation was then generalized (at the suggestion of E. L.
Titlebaum) to include wideband signals that apply to animal sonar
(Altes, 1971a; 1971b).

The SIR approach has recently been further extended by
M. Decouvelaère (this volume). Another extension, in which the
point target assumption is replaced by a distributed target with
known impulse response, has been considered by several authors
(Harger, 1970; Lee and Uhran, 1973; Altes and Chestnut, unpublished).

A problem with the SIR formulation is that it is not specifically adaptive, i.e., it does not specify a method for updating
the clutter or reverberation description and/or the estimate of
target impulse response. Recent adaptive detector formulations
(Picinbono, 1978; Brennan and Reed, 1973) allow the receiver configuration to change with the environment when the transmitted
waveform is fixed. When the signal is allowed to vary, however,
adaptive estimation of the clutter statistics (which are signal
dependent) becomes more complicated.

Bel'kovich and Dubrovskiy (1976) have argued that an adaptive
receiver without an adaptive waveform generator may be a reasonable
model for dolphin echolocation. Au, et al. (1974), however, have
found that Tursiops seems to use unusually high frequencies for
long range detection, apparently to minimize the effect of

low-frequency noise. Higher frequencies also result in a narrower
beam width, which implies spatial as well as temporal discrimination
against omnidirectional low-frequency interference.

Dziedzic and Alcuri (1977) have observed a target-dependent
signal change at close range in Tursiops. It would seem that waveform changes at close range should be associated with target characterization behavior, e.g., estimation of target impulse response
in the presence of clutter and noise (Altes, 1977), rather than
with a detection method such as SIR maximization.

III. CORRELATION PROCESSING FOR CHARACTERIZATION OF
      RANGE-EXTENDED TARGETS

The echo from a range-extended target can be parameterized
in many ways. A standard method for analyzing such an echo is to
correlate it with a sequence of orthonormal functions (Fig. 3).
When echo arrival time is unknown, the correlation operations can
be performed with linear filters (bandpass filters for Fourier
series analysis), and the sum of the squares of the filter outputs
(noncoherent detector response) can be used to detect the target
echo and to generate a sampling time, as shown in Fig. 3. The
desired parameters or expansion coefficients are the sampled filter
outputs, where samples are taken at the instant when the sum of
the squared responses is maximized. A matched filter for a known
signal can be synthesized by multiplying the output of each orthonormal filter by an appropriate coefficient and by summing the
weighted filter outputs (coherent detection). The process can be
made adaptive by using a weighted sum of coherent and noncoherent
detector outputs, where the weights depend upon the receiver's
uncertainty about the appropriate coefficients to use in the
coherent detector (Glaser, 1961).

Many different sets of orthonormal basis functions can be
used; sines and cosines are only one example. Which of these sets
is most relevant to animal echolocation? In fact, why should animals be restricted to an orthogonal basis set? Splines (Greville,
1969) and linear prediction coefficients (Makhoul, 1975), for
example, are useful non-orthogonal descriptions of a time function. An even more basic question is: Why should the echo be
parameterized at all? Why not just perform detection and classification with the echo waveform itself?

These questions will be considered in the following order:
First, "why parameterize?" Second, "which parameters or basis
functions are most relevant?" Third, an example will illustrate
the importance of orthogonality.

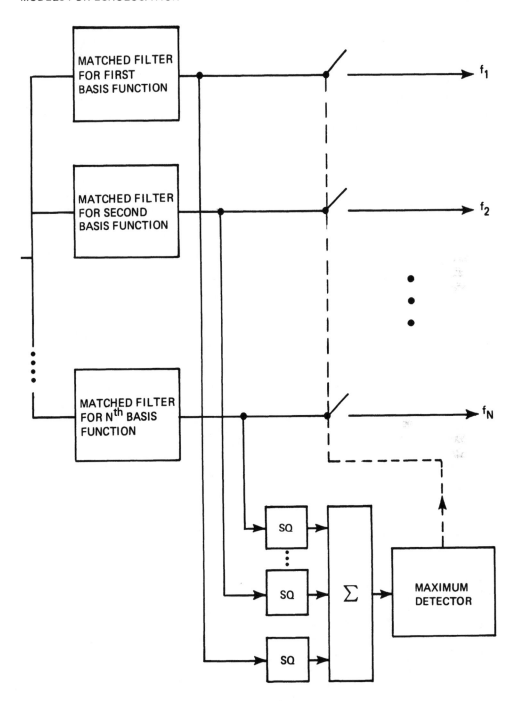

Fig. 3. Orthogonal decomposition of a signal with unknown
waveform and unknown epoch. SQ - square law device.
See Glaser (1961).

Parameterization can be justified for biological systems by
equating the word "parameter" with the word "feature." Many experiments indicate that feature extraction is performed in animal echolocation. Bullock et al. (1968) found neural populations that were
sensitive to rise time, intensity, frequency, frequency modulation,
and sound direction in four dolphin species. Suga (1972) has
described similar feature sensitivity in single bat neurons. In
man-made systems, efficient parameterization lowers memory requirements and reduces the number of operations that are required for
signal classification.

Given that echo parameterization is a reasonable operation
for a biological system, what parameters or features are most meaningful or important for survival? In the study of echolocation,
at least two different approaches to this problem have been pursued.
One approach is to start with scattering theory and to obtain a set
of parameters that are theoretically important for distinguishing
between different sonar reflectors. Another approach is to
empirically determine the important features for target classification by using pattern recognition methods. The first approach will
be described in the following paragraphs, and the second method
will be described in Part IV.

Research on the prediction of target scattering characteristics
has resulted in two main theoretical approaches. Geometrical and
physical optics have evolved into a technique that is known as the
geometrical theory of diffraction (Bechtel, 1976). This theory
involves the prediction of scattering characteristics from separate
structures such as the tip of a cone or an edge, and it includes
creeping wave effects (i.e., waves that propagate around the surface of the object). The echo components from various target structures are combined with proper delay, in order to form the composite
echo. The geometrical theory of diffraction is primarily a high
frequency approach for wavelengths that are smaller than the target
(Kouyoumjian, 1965).

The second approach to target scattering prediction is the
use of natural modes or resonances (Miller, et al., 1975). Natural
modes tend to dominate the trailing edge of an echo, whereas specular components (mirror-like reflections or glints) with higher
amplitudes are found at the beginning of the echo. The advantage
of natural mode characterization is that the resonant frequencies
are independent of aspect, and it would thus seem that the target
echo need not be modelled as an aspect-dependent random process.
Experimental results with various sonar targets, however, have
shown that, although natural resonances may be independent of
aspect, the excitation of a given resonance and the energy that is
re-radiated toward the receiver is, in fact, aspect-dependent.
Furthermore, the trailing edge of the echo tends to be attenuated

relative to the specular parts of the echo. Finally, the trailing edge of an echo is difficult to analyze if the target is immersed in clutter.

According to Bechtel and Ross (1966), the geometrical theory of diffraction predicts that many scattering centers will have cross-sections that vary as λ^n, where λ is wavelength and $-2 \leqslant n \leqslant 2$. If cross section varies as λ^n, then the associated transfer function must vary as a half-power of frequency, and the echo spectrum is $(j\omega)^{n/2}U(\omega)$ for $-2 \leqslant n \leqslant 2$, where $U(\omega)$ is the Fourier transform of the transmitted signal. An idealized echo from a scattering center or highlight can be synthesized by passing an impulse through a filter with transfer function $(j\omega)^{n/2}U(\omega)$. The parameter n is range dependent, and changes from one highlight to the next. A highlight causes a local maximum in the echo envelope, and in Part II it was assumed that all highlights corresponded to transfer functions which varied as $(j\omega)^0 = 1$. Each scattering center or highlight echo was thus assumed to be a delayed version of the transmitted signal.

In 1975, an attempt was made to generalize the point target highlight model by applying a spline-like representation to measured target impulse responses (Altes and Reese, 1975; Altes, 1976b). Scattering centers in the spline model were assumed to have transfer functions $(j\omega)^n$ rather than $(j\omega)^{n/2}$. This difference has little effect upon sonar receiver configuration, and the following discussion of the spline model will use half-powers of $(j\omega)$.

The receiver consists of a detector/echo parameter estimator that is similar to the one in Fig. 3. The frequency domain echo from the mth scattering center can be written

$$E_m(\omega) = \sum_{n=-2}^{2} f_{nm}(j\omega)^{n/2} U(\omega) \exp(-j\omega\tau_m) \qquad (7)$$

where τ_m is the delay associated with highlight range, $U(\omega)$ is the Fourier transform of the transmitted signal, and f_{nm} is an unknown parameter. The problem is to estimate the parameters f_{nm}, $n=-2$, -1, 0, 1, 2, and the delay τ_m for each highlight.

For the echo model in (7), the basis functions for echo analysis are the functions $(j\omega)^{n/2}U(\omega)$, and the matched filters in Fig. 3 have transfer functions $[(j\omega)^{n/2}U(\omega)]*$, $n=-2$, -1, 0, 1, 2. Since the basis functions are not orthogonal, minimum mean-square error estimates of the parameters $\{f_{nm}\}$ are obtained by passing the filter outputs through a matrix transformation that compensates for this lack of orthogonality. This transformation can be implemented by a set of interconnections between the filter outputs in Fig. 3.

One additional assumption can be used to completely specify
the filter transfer functions and the signal spectrum $U(\omega)$. Assume
that the filters with transfer functions $[(j\omega)^{n/2}U(\omega)]*$, $n=-2,\ldots,2$
are all constrained to have the same time-bandwidth product. Time-
bandwidth product is a measure of filter complexity, and a constraint on the product of impulse response duration and transfer
function bandwidth is equivalent to a system complexity constraint.
Time-bandwidth product can be kept invariant by requiring all the
filter transfer functions to be scaled versions of one another, e.g.,

$$(j\omega)^{n/2}\ U(\omega)\ \propto U(\omega/k^{n/2}) \ . \tag{8}$$

For $k > 1$, the filter for $n=1$ is broader in the frequency domain
than the filter for $n=0$, but its impulse response is narrower in the
time domain. Equation (8) can be solved for $U(\omega)$. The solutions
correspond to linear-period-modulated signals that can be made to
resemble either bat pulses or dolphin echolocation clicks (Altes,
1976b). The constraint on time-bandwidth product can also be
related to a condition for Doppler tolerance (Altes and Reese,
1975; Altes and Titlebaum, 1970).

If all the filter transfer functions $U*(\omega/k^{n/2})$, $n=-2,\ldots,2$,
are scaled versions of one another, then they all have the same
ratio of center frequency to bandwidth, i.e., the same Q. The
filter bank in Fig. 3 then becomes a set of overlapping proportional
bandwidth filters, as in models of the peripheral auditory system.

The above theoretical construction seems to exhibit the kind
of "inner beauty" and self-consistency that was expressed by Dirac
and Feinman, as quoted in the introduction. Starting with scattering theory ideas, one obtains mathematical approximations for bat
and dolphin echolocation signals and an estimator/detector that is
similar to a cochlear spectrum analyzer. The theory seems to apply
to vision as well as to echolocation (Altes, 1976a; Hauske, Wolf,
and Lupp, 1976) and perhaps even to electrolocation in Mormyrids.

Unfortunately, the beauty now appears to be only skin deep.
There are two main flaws in the theory, both of which provide important lessons for future modelling efforts:

(i) Non-separable highlights.

The theory implicitly assumes that the echo from each scattering center is not contaminated by reflections from other parts of
the target. Computer processing of actual sonar echoes, however,
seems to indicate that a highlight is rarely isolated in range
(Skinner, Altes, and Jones, 1977). Sonar highlights are caused by
range-distributed impedance changes, and a large highlight is often
surrounded by many smaller highlights. The combined echoes from

these small highlights have an effect upon the receiver that is
similar to noise, and when environmental noise is added, estimates
of the parameters f_{nm}, n=-2, -1, 0, 1, 2, become unreliable.

(ii) Non-orthogonality.

     It is not immediately obvious that a model for animal echo-
location should employ orthogonal basis functions. The use of non-
orthogonal basis functions, however, can severely degrade the performance of an estimator in a noisy environment. This effect is
theoretically predicted by the matrix Cramér-Rao bound (Van Trees,
1968), which has been applied to the jointly estimated parameters
f_{nm}, n=-2,...,2 by W. Hodgkiss (unpublished). Lower bounds on the
variances of the estimated parameters are given by the diagonal
elements of a matrix J^{-1}. The elements of J are proportional to the
inner products (i.e., the degree of non-orthogonality or overlap)
between basis functions. J is called the Fisher information matrix.
For the transfer functions $U(\omega/k^{n/2})$ that are obtained from (8), the
diagonal elements of J depend upon the areas under each of the
curves in Fig. 4, and the off-diagonal elements depend upon the
areas of overlap between the different functions. From Fig. 4, we
see that the off-diagonal elements can be nearly as large as the
diagonal elements. This observation means that the elements of
J^{-1} can be very large, and estimation errors can be considerable
even for a low level of environmental noise.

     The matrix that is used to transform the filter outputs in
Fig. 3, in order to estimate the parameters f_{nm} , is proportional
to J^{-1}. The large off-diagonal elements of J make J^{-1} difficult to

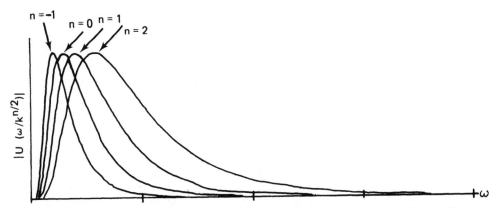

     Fig. 4. Filter transfer functions used for estimation and
detection of target parameters, shown with linear frequency and
amplitude scales. On a log-log scale, the functions become a
sequence of identical parabolas, uniformly spaced in logω.

implement, i.e., the elements of J^{-1} are large, and they must be exactly specified (with double precision computer arithmetic) or the system will not work. In mathematical terms, J is ill-conditioned (Skinner, Altes, and Jones, 1977). If orthogonal basis functions were used, J would be an identity matrix, and the receiver would be as shown in Fig. 3.

There may be some untried modifications of the system that will make it perform better. The use of an iterative subtraction method may help to separate highlights (Skinner, Altes, and Jones, 1977). An assumption that only one of the coefficients f_{nm}, n=2,...,2 is nonzero for each highlight may provide a way to circumvent the ill-conditioned matrix transformation J^{-1}. The estimator in this case can be a maximum likelihood hypothesis test, and only the maximum filter response is used.

In any case, the discussion has illustrated the power of a model, combined with computer processing of actual data, to test ideas and to reveal mistakes in assumptions. Insight into the target parameterization problem that is encountered in echolocation has also been obtained. Highlights are often difficult to analyze as separate echo waveforms, and orthogonal (or nearly orthogonal) basis functions provide better estimates of echo parameters in the presence of noise than can be obtained with non-orthogonal functions.

IV.   SCATTERING FUNCTIONS, TIME-FREQUENCY ENERGY DENSITY FUNCTIONS, AND SPECTROGRAM CORRELATION

Another representation for a sonar target is the target's scattering function (Van Trees, 1971). The scattering function, $\bar{S}(t,f)$, describes the expected extent to which a sonar target (or time-varying communication channel) spreads out an input signal in time and frequency. By using a channel input-output relation derived by Bello (1963) and some spectrogram properties obtained by Ackroyd (1971), the spectrogram of a target echo $S_{zv}(t,f)$ can be related to the spectrogram of the input signal $S_{uv}(t,f)$ by the equation (Altes, 1978b)

$$E\left\{S_{zv}(t,f)\right\} = \bar{S}(t,f) * E\left\{S_{uv}(t,f)\right\} \tag{9}$$

where (*) denotes a two-dimensional convolution operation in time and frequency, and $E\{\cdot\}$ denotes expected value or ensemble average. The spectrogram of the echo is thus a smeared version of the signal spectrogram, where the degree of smear is determined by the scattering function of the target.

A scattering function is useful for characterizing targets with time-varying transfer functions, such as flying insects. If intra-pulse time variations are neglected, then a time-frequency description of a target with transfer function H(f) and impulse response h(t) is given by the time-frequency energy density function $e_{hh}(t,f)$ of the target impulse response (Rihaczek, 1968), where

$$e_{hh}(t,f) \equiv h(t) \, H^*(f) \, \exp(-j2\pi ft) \, . \tag{10}$$

It can be shown (Altes, 1979b) that for a time-invariant target, the spectrogram of a target echo $S_{zv}(t,f)$ can be written

$$S_{zv}(t,f) = e_{hh}(t,f) \overset{t}{*} S_{uv}(t,f) \tag{11}$$

where $S_{uv}(t,f)$ is again the spectrogram of the input signal and where $\overset{t}{*}$ denotes one-dimensional convolution or smearing in the time direction. Eq. (11) implies that the target time-frequency energy density function can be estimated with a signal that has large bandwidth and short duration, such as a dolphin echolocation pulse.

The time-frequency energy density function provides a useful description of a target because it contains information about both highlight distribution and energy spectrum. Local maxima of $|h(t)|^2$, the time envelope of the target impulse response, provide highlight locations. Using (10), one can demonstrate that

$$|h(t)|^2 = \int_{-\infty}^{\infty} e_{hh}(t,f) \, df \tag{12}$$

and highlight ranges can be obtained from $e_{hh}(t,f)$. Similarly,

$$|H(f)|^2 = \int_{-\infty}^{\infty} e_{hh}(t,f) \, dt \tag{13}$$

where $|H(f)|^2$ is the energy spectrum of the target impulse response. Utilization of the energy spectrum for target classification is discussed in the section by R. A. Johnson.

Now suppose that the scattering function or the time-frequency energy density function of a target is known, and we wish to detect the target when its echo is added to white, Gaussian noise. From (9) and (11), knowledge of the target scattering function or t-f energy density function is equivalent to knowledge of the expected echo spectrogram. The locally optimum detector in this case is a spectrogram correlator (Altes, 1978b), i.e., a device that computes the quantity

$$\ell(\underline{z}) = \sum_{i=1}^{N} \sum_{j=1}^{M} z_{ij} \, E\{s_{ij}\} \tag{14}$$

where z_{ij} are samples of the data spectrogram and $E\{s_{ij}\}$ are samples of the expected spectrogram of the echo waveform.

A spectrogram correlator model for echolocation is especially attractive, since the mammalian peripheral auditory system functions like a bank of bandpass filters followed by envelope detectors (Siebert, 1968, 1970; Evans, 1977; de Boer, 1975). The resulting data representation is very similar to a spectrogram.

The spectrogram correlator can be used for classification as well as for detection by finding the maximum correlator response for a number of different reference spectrograms. This procedure can be very time consuming, however, since the number of samples NM in (14) is generally large. It would be more efficient to obtain meaningful features for distinguishing the spectrograms of different targets, and to decide whether or not these features are present in the data spectrogram.

An attempt to predict features from scattering theory, and to use non-orthogonal basis functions for feature extraction, has already been described. In the following analysis, features that are empirically determined from the data, along with orthogonal basis functions, will be used.

The desired number of features is limited in practice. A limited set of features is associated with a set of basis functions that is incomplete; one cannot exactly reconstruct every possible data waveform with a limited number of basis functions. For a given number of feature vectors (orthogonal filters) N, one can obtain a "best" incomplete N-dimensional basis set. A "best" basis corresponds to the N orthogonal functions that give the least average squared error between an echo that is reconstructed from N coefficients and the corresponding data echo. The average is taken over a representative set of data echoes. The resulting set of N feature vectors are called the "principal components" corresponding to the representative set of data echoes (Chien and Fu, 1968).

Since a limited number of basis functions are used to convey the maximum possible information about the data, one can assume (in the absence of a better approach) that these basis functions correspond to the most important features for classification.

In order to obtain a representative set of data, echoes of wideband, dolphin-like signals were obtained from targets with different shape, size, and composition, and for many aspects of a solid aluminum cylinder, from end-on to broadside. Two-dimensional principal components were then determined from the echo spectrograms. The resulting orthogonal basis functions represent the most important features for spectrogram echo classification. Seven of these components were used in a target classification experiment that will be described. The three most important components are shown in Fig. 5 (Altes and Faust, 1978).

The functions in Fig. 5 have interesting implications for models of the acoustic and visual systems. Suppose that each of the features in Fig. 5 corresponds to a neuron in the auditory cortex. The neuron that is associated with Fig. 5b, for example, has an inhibitory center at 40 kHz, surrounded by an excitatory area in frequency and time. The neuron in Fig. 5c is excited by spectrogram outputs that appear early in time, and it is inhibited by outputs that appear later in time. Such neurons display not only lateral inhibition for neural "sharpening" in the frequency domain (Nilsson, 1978), but also forward and/or backward masking in the time domain (Sparks, 1976). The similarity to receptive fields in vision (Hubel and Wiesel, 1968) is obvious. Our results suggest that visual neurons with specialized receptive fields may form an orthonormal basis for pattern recognition, and the well-known excitatory center/inhibitory surround cell is one of the principal components for such a basis.

The direction of the ridges in Fig. 5 seems to indicate that temporal variations (highlight locations) are more important than frequency variations (energy spectra) for classification. This effect, however, is strongly dependent upon the spectrogram time window (the impulse response duration of the filters that are used to construct the spectrogram) and upon the "representative" set of data echoes that are used to obtain the principal components.

Vel'min and Dubrovskiy (1976) have found a "critical interval" in dolphin audition. If a time-limited signal and time-limited interference are separated in time by more than a critical interval, the interference does not affect signal detectability. Just as a critical band can be used to model the frequency window or bandwidth of filters that are used to construct a spectrogram (Johnson, 1968b) the critical interval can be used to model the duration of the time window for spectrogram analysis. According to Vel'min and Dubrovskiy, the critical interval is 200-300 µsec long. The time window corresponding to Fig. 5 is 128 µsec long.

Before principal component analysis occurs, a data spectrogram must be registered in time, i.e., the temporal position of the data

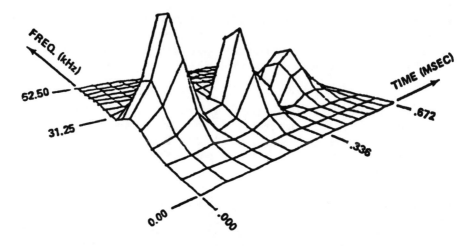

TOP VIEW

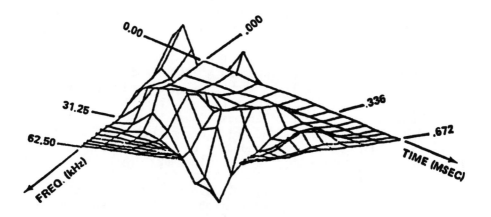

BOTTOM VIEW

Fig. 5a. Principal components or basis functions for the analysis of echo spectrograms.

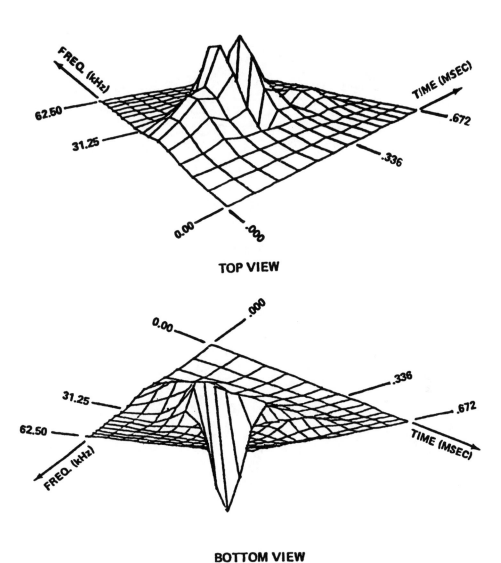

TOP VIEW

BOTTOM VIEW

Fig. 5b. Principal components or basis functions for the
analysis of echo spectrograms.

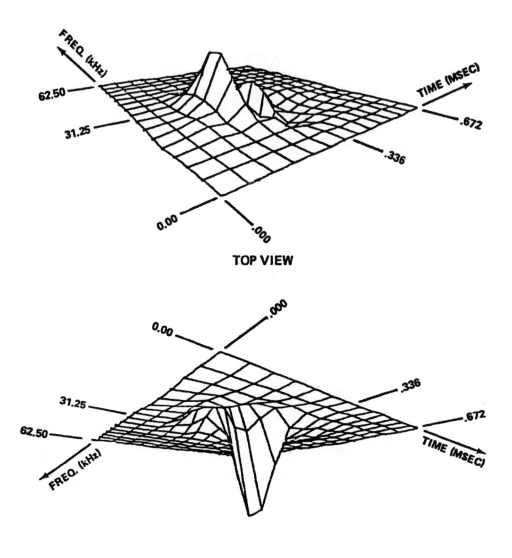

TOP VIEW

BOTTOM VIEW

Fig. 5c.   Principal components or basis functions for the
analysis of echo spectrograms.

relative to the basis set must be determined. In the computer algorithm, this registration was accomplished by a noncoherent energy detector as in Fig. 3. Data energy was measured in a 128 μsec long time interval, and the energy in the interval as a function of interval starting time (the left-hand edge of the interval) was plotted. A maximum in the resulting energy versus time function was used to register the data spectrogram relative to the reference spectrograms. This registration process can be generalized by using Glaser's formulation (Glaser, 1961), but impressive classification performance can be obtained by using energy versus time as the only method of registration.

When the expected spectrogram of an echo can be predicted, a locally optimum echo detector is a spectrogram correlator. This detector implements a test for deciding between two alternative hypotheses, H_1 and H_0, where H_1 is the hypothesis that a target with specified scattering function plus white, Gaussian noise is present, and H_0 is the hypothesis that only noise is present. A classification problem can be viewed as a multiple hypothesis test that uses the same basic operations as the two-hypothesis case. For K different targets, a maximum likelihood classifier forms K two-hypothesis tests that decide between H_k and H_0, k = 1, 2, ..., K. The test statistics (in this case, the spectrogram correlator responses) for the K target hypotheses are compared, and the largest statistic is chosen. If this largest statistic exceeds a threshold, then the receiver decides that the corresponding target is present. If the largest correlator response is too small, then the receiver decides that only noise is present.

Principal component analysis does not change the basic structure of the classifier. Correlation between data spectrograms and reference spectrograms is performed in a new coordinate system with reduced dimensionality. Simplification occurs because of this reduced dimensionality, and because the reference functions for spectrogram correlation are the same for all hypotheses (the same features are extracted, regardless of hypothesis). When the coefficients of data and reference spectrograms in the new coordinate system are energy normalized, a nearest neighbor classifier is equivalent to a spectrogram correlator. A nearest neighbor classifier measures the Euclidean distance between a data point (i.e., a representation of the data spectrogram) and each reference point (i.e., a representation of each reference spectrogram). The reference point that is closest to the data point determines the classification of the data spectrogram. If there were three principal components, data points and reference points would appear in an ordinary three-dimensional x,y,z coordinate system, and energy normalization would cause all the points to fall on a sphere centered at the origin. The intersection of the sphere with the x, y, and z axes would correspond to the patterns in Figs. 5a, 5b, and 5c, respectively.

    For the classification experiment, half the echoes in the
data set were obtained from a 6.35 cm x 17.8 cm aluminum cylinder
that was rotated from end-on aspect to broadside. The other half
of the echoes were from various other objects such as hollow,
open-ended metal cylinders with various wall thickness, styrofoam
cubes and cylinders, a metal sphere, etc. An interesting classification problem was to distinguish the solid aluminum cylinder from
the rest of the targets in a white noise background. Other problems
were to distinguish cylinders from non-cylinders and metal targets
from non-metal ones.

    Reference vectors from all the targets were used in a nearest-
neighbor classifier with seven principal components. No attempt
was made to generalize the classifier by dividing the seven-
dimensional principal component space into decision regions. Such
a division is straightforward; a nearest-neighbor classifier is
replaced by a least-squares minimum distance classifier (Ahmed and
Rao, 1975). The least-squares classifier represents a method for
classifying future data on the basis of past experience, and it
constitutes a model for learning and generalization. An adaptive
version of such a classifier was first derived as a model for biological pattern recognition (Widrow, 1962).

    Classification performance of a spectrogram correlator with
seven principal components is shown in Figs. 6a-6c. In these figures, chance classification (i.e., the result of random guessing)
occurs at different levels because of different numbers of target
echoes in each class. SNR is echo energy divided by average noise
power at the output of a bandpass filter with a 90 kHz wide bandwidth. This performance is compared with some suboptimum techniques
in Altes and Faust (1978).

    In summary, it has been found that an empirical determination
of target features, combined with an orthogonal component analysis,
leads to impressive classification performance. The empirically
determined feature vectors are patterns on a time-frequency plane.
The correlation of these feature vectors with data spectrograms,
combined with nearest-neighbor classification, is a form of spectrogram correlation process. Spectrogram correlation is the basis
of a locally optimum test between a signal with known expected
spectrogram and white, Gaussian noise, so it is not surprising that
the classifier performs well in additive Gaussian noise. A signal
with known expected spectrogram corresponds to an echo from a target
with known scattering function or time-frequency energy density
function. The t-f energy density function can be related to the
highlight structure of the reflector and also to the energy spectrum of the target echo.

    The analysis has illustrated a strong interconnection between
detection, estimation, and classification. If the data is classified

a. Pr[misclassification, 2.5 x 7" Al. Cyl. versus other targets]

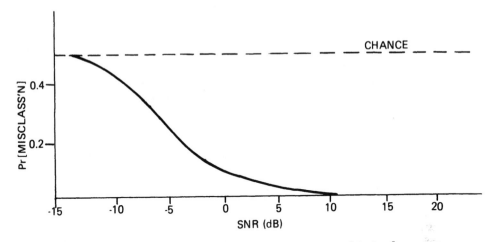

b. Pr[misclassification, cylinder versus non-cylinder]

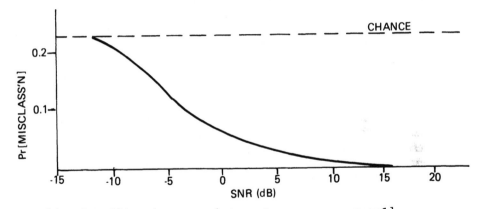

c. Pr[misclassification, metal target versus non-metal]

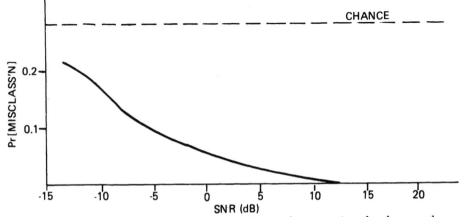

Fig. 6. Classification performance in a white noise background.

as noise, then it is rejected, and the <u>classifier</u> has functioned as a <u>detector</u>. The coefficients that are extracted by correlation of spectrograms with principal component feature vectors are minimum mean-square error estimates of the feature values, and the classifier/detector utilizes a parameter <u>estimator</u>. The relation between detection, estimation, and classification is becoming an increasingly popular theme in statistical communication theory (Bangs and Schultheiss, 1973; Scharf and Nolte, 1977).

A spectrogram processor is an attractive model for echoloca-tion because the encoding of sound by the peripheral auditory system is similar to the formation of a spectrogram, and because the underlying target model (the scattering function) takes account of time-varying transfer functions. The next part of this review considers behavioral tests to determine whether spectrogram corre-lation is a viable model for echolocation.

V.   PERFORMANCE OF SPECTROGRAM PROCESSORS, AND TESTS FOR THEIR UTILIZATION IN ANIMAL ECHLOCATION

The following paragraphs describe quantitative characteristics of a spectrogram correlator model, tests that exploit these char-acteristics to determine the validity of the model, and the outcomes of some of these tests on human subjects.

If a spectrogram correlator is to perform as well as a matched filter, signal-to-noise ratio should be increased by approximately 5 dB (Altes, 1978b, 1979b). A five dB difference in detection per-formance, however, is difficult to measure experimentally, since internal system noise is generally unknown.

For a two-alternative, forced choice (2AFC) experiment, the probability of a correct choice is a monotone increasing function of SNR, and this function is significantly steeper for a matched filter than for an energy detector (Altes, 1978b). Data for a 2AFC experiment have been obtained for human subjects, using a sine wave signal with a rectangular envelope. For such a signal, a spec-trogram correlator is equivalent to an energy detector. The results of this experiment (Green and Swets, 1966) favor a spectrogram cor-relator or energy detector over a matched filter.

For a sum of K sinusoids in white noise, the SNR at the output of a spectrogram correlator is

$$\text{SNR(SC)} \propto \sum_{k=1}^{K} E_k^2 \, / \, [\sigma^2 \sum_{k=1}^{K} E_k] \tag{15}$$

where σ^2 is the average noise power and E_k is the energy of the kth sinusoid. For a matched filter,

$$\text{SNR(MF)} \propto \sum_{k=1}^{K} E_k / \sigma^2 . \tag{16}$$

For a spectrogram correlator, signal detectability should vary as the sum of the squared energies of the sinusoidal components, if the signal is energy normalized. This behavior is similar to Green's summation law, which describes human detection of a sum of sinusoids (Green, 1958; Green, McKey, and Licklider, 1959). Energy normalization can be approximated by an automatic gain control (AGC), a function that may be accomplished by middle ear muscle contraction.

The frequency of a narrowband sinusoid with duration T can be estimated with a spectrogram correlator. The standard deviation or uncertainty Δf for the frequency measurement is proportional to $T^{-\frac{1}{2}}$ (Altes, 1978b, 1979b). For a matched filter, Δf is proportional to T^{-1} (Cook and Bernfeld, 1967). A number of experiments involving frequency discrimination in humans have been summarized by Siebert (1970). In Siebert's empirically determined equation, Δf varies as $T^{-\frac{1}{2}}$.

The arrival time of a wideband pulse with bandwidth B can also be estimated with a spectrogram correlator. The standard deviation or uncertainty $\Delta \tau$ for this epoch measurement is proportional to $B^{-\frac{1}{2}}$ (Altes, 1978b, 1979b). For a matched filter, $\Delta \tau$ is proportional to B^{-1} (Cook and Bernfeld, 1967).

Predictions of detection and estimation performance for a spectrogram correlator seem to be consistent with data from a number of behavioral experiments performed with human subjects. Data from similar experiments with echolocating animals have apparently not yet been compared with the predicted performance of a spectrogram correlator model. The predictions in the foregoing paragraphs, along with some of the techniques that have been used for human experimentation, provide a fertile area for future behavioral work with bats and dolphins. These quantitative predictions illustrate how a model (or two competing models) can produce a feedback mechanism for experimentation. Future experiments are obtained from models that are based upon past data.

VI.   BINAURAL DETECTION AND LOCALIZATION

a.   Theoretical Direction (Delay Vector) Estimators, and
      Associated Detector Configurations

        Recent experiments that involve sound localization in echoloca-
tion will be described, but first it is worthwhile to examine theoretically optimum direction estimators and detectors that have
been developed for man-made radar/sonar systems. The configuration
of these estimators and detectors depends upon knowledge of the
echo waveform. We shall consider the two extreme cases: (i) echo
waveform unknown, (ii) echo waveform known exactly.

        (i) Echo waveform unknown.

        When the echo waveform that impinges upon a series of iso-
tropic array elements is unknown, one can infer the direction of
the echo source from an estimate of the delay vector, i.e., the
delays in arrival time between each of the elements and the first
element. Hahn and Tretter (1973) and Hahn (1975) have shown that a
maximum likelihood estimate of the delay vector is obtained by pair-
wise cross-correlation of the signals that emerge from the different
array elements. For a binaural system, an ideal interaural delay
estimate involves cross-correlation of the signals from each ear,
when the echo waveform is unknown. The cross-correlator response
can be used for detection as well as for angle estimation (Wolff,
Thomas, and Williams, 1962).

        (ii) Echo waveform known exactly.

        When the echo waveform is known exactly, an ideal receiver
should synthesize a matched filter at the output of each array element. When the elements are anisotropic due to head shading, pinna
shape in bats, or multiple sound paths in the dolphin head, the
matched filters must incorporate a direction hypothesis. For each
direction hypothesis, a different matched filter is used. Monaural
direction hypotheses are also required when the transmitted signal
is direction-dependent, as in Tursiops (Evans, 1973). This complexity is offset by accurate monaural localization capability, which
exploits echo waveform and amplitude changes that are induced by
a direction change (Livshits, 1974; Altes, 1978a). A binaural
direction hypothesis is implemented by using the appropriate matched
filter for each ear and by summing delayed matched filter responses,
where the delay is determined from the direction hypothesis. The
resulting sum is a matched filter in space and time. The direction
hypothesis that gives the maximum space-time matched filter response
corresponds to a maximum likelihood direction estimate (Altes,
1978a). This filter response can also be used for estimation of echo
amplitude (Davis, Brennan, and Reed, 1976).

A sequence of space-time matched filters with different direction hypotheses can be used for detection as well as for estimation. To detect a waveform that is known except for direction, the likelihood ratio for detection of an echo from a particular direction is multiplied by the probability density function that describes the direction uncertainty of the receiver, and the product is integrated over all possible directions. To write this detection statistic in the form of an equation, we let a column vector $\underline{r}$ denote the sampled data from both ears, and we let $\underline{m}(\theta)$ be a noise-free version of $\underline{r}$ when an echo is received from direction θ. It follows that $\underline{r}^T\underline{m}(\theta)$ is the response of a space-time matched filter to a binaural data vector $\underline{r}$, and $\underline{m}^T(\theta)\underline{m}(\theta)$ is the response of the filter to a noise-free echo from direction θ. In Gaussian noise with variance σ^2, the detection statistic is

$$\ell(\underline{r}) = \int_{-\pi}^{\pi} \exp\{[\underline{r}^T\underline{m}(\theta) - \frac{1}{2}\underline{m}^T(\theta)\,\underline{m}(\theta)]/\sigma^2\}\,p(\theta)\,d\theta \quad . \qquad (17)$$

According to (17), the signal is detected by forming a monotone function of the matched filter response $\underline{r}^T\underline{m}(\theta)$ for each angle hypothesis θ, and by summing a weighted version of this function over all θ-values, where the weighting function is a probability distribution that describes the receiver's uncertainty concerning θ. An adaptive estimator/detector can be implemented by updating $p(\theta)$ on the basis of accumulated echo data (Nolte and Hodgkiss, 1975). When θ is known exactly, e.g., $\theta = \theta_o$, then $p(\theta) = \delta(\theta - \theta_o)$, and $\ell(\underline{r}) \propto \exp[\underline{r}^T\underline{m}(\theta_o)]$, the response of a filter that is matched to both signal waveform and direction.

For a binaural system, an alternate form of maximum likelihood direction estimate for a known signal is obtained by subtracting the delayed matched filter responses from the two ears. If the direction hypothesis is correct and if the matched filter response of each ear is multiplied by an appropriate gain, a null will be observed at the subtractor output in the absence of noise. The subtractor implementation results in a "difference beam" at the receiver (Davis, Brennan, and Reed, 1976), and the sign of a non-zero subtractor response can be used to track the target. A similar estimation/tracking philosophy for delay, rather than direction, results in a delay-lock discriminator (Spilker and Magill, 1961). The delay-lock discriminator uses a time derivative of the reference signal, and the subtractor or difference beam uses a spatial derivative of the reference signal (Davis, Brennan, and Reed, 1976).

b. Models for Binaural Interaction

A detailed study of binaural interaction in humans has been
undertaken by H. S. Colburn and his associates (Colburn, 1973,
1977; Colburn and Latimer, 1978; Stern and Colburn, 1978). The
resulting model is shown in Fig. 7. According to this model, corresponding auditory nerve fibers from each ear (i.e., fibers tuned
to the same characteristic frequency) are combined after an internal,
interaural delay τ_m is applied to the output of one of the ears.
The combining operation is a coincidence counter for neural firings,
which depends upon a window function $f(t_i-t_j-\tau_m)$, where t_i-t_j is
the difference between firing times for the corresponding nerve
fibers. The coincidence counts L_m are multiplied by weights c_m
that depend upon interaural pressure differences. The weighted
sum of the coincidence counter responses is L_B, a binaural decision
variable that can be used for localization. Colburn has pointed
out that this model is similar to the models that were obtained
by Jeffress (1948), Sayers and Cherry (1957), and Licklider (1959).

The model in Fig. 7 is similar to an ideal angle estimator/
detector for a signal with unknown waveform. A coincidence counter
seems to be a neural analog to a polarity coincidence correlator
(Wolff, Thomas, and Williams, 1962) or a dead-zone limiter coincidence correlator (Kassam and Thomas, 1976, 1977) if neural firings
are correlated with threshold crossings of an input signal. At
higher frequencies, however, neural firings may represent average
signal amplitude over a given time interval. Since spontaneous
action potentials can occur in the absence of an input signal, a
coincidence counter could then implement a summation operation. In
this case, the model resembles a spatial matched filter for a
known echo waveform.

An ambiguity in the mathematical significance of a coincidence
counter can thus be associated with an ambiguous interpretation of
the corresponding estimation/detection problem. Could this ambiguity
be exploited by the biological system? By a subtle change in neural
coding or in window function operation, it may be possible to adapt
the behavior of the system to a particular estimation/detection
problem (waveform known, unknown, or partially known).

c. Neurophysiological Data

If any localization system in nature fits into the "signal
with known waveform" category, it is the processor for the constant
frequency echo component of the mustache bat (Pteronotus parnellii
rubiginosus). The CF component of the mustache bat's signal is carefully tailored to furnish the receiver with a predictable waveform.
When the input to a linear filter is a sine wave, the output is a
sine wave at the same frequency (usually with different amplitude

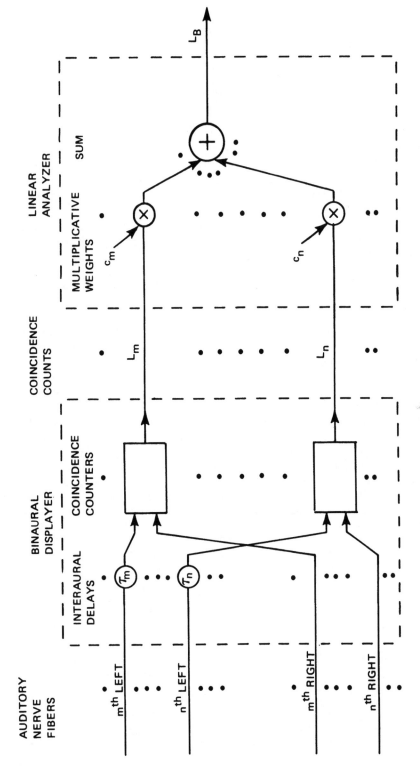

Fig. 7.   Generation of binaural decision variable L_B from auditory-nerve firing patterns (From Colburn and Latimer, 1978).

and phase), so echoes from motionless targets are similar to the
transmitted signal. (This fact helps to explain the compatibility
of the point target assumption with a narrowband radar/sonar system.)
In Pteronotus, Doppler shift at the receiver is minimized by adjust-
ing the transmitted signal frequency in order to compensate for
range rate. Similar behavior has been observed in Rhinolophus
(Schnitzler, 1973).

Menabe, Suga, and Ostwald (1978) have found that the CF processing area in the auditory cortex of the mustache bat has two
functional subdivisions. One population of neurons (the E-E neurons)
is excited by tone-burst stimulation of either ear, while another
population (the I-E neurons) are inhibited by stimulation of one
ear and excited by stimulation of the other. When both ears are
stimulated simultaneously by a sound source at various azimuth
angles, the responses of the E-E neurons are found to be angle
insensitive, while the I-E neurons are sensitive to angle. The
responses are also amplitude dependent. Most of the E-E neurons
respond best to weak tone bursts, while the I-E neurons respond
best to stronger stimuli.

Menabe, et al., have speculated that the I-E neurons are used
for accurate localization of echoes from targets at short range,
while the E-E neurons are used for long-range detection.

From (17), a direction-insensitive detector is ideal when the
target direction is not well known, i.e., when $p(\theta)$ is broad. As
more echo data becomes available, the detector should become more
angle sensitive, so that detection performance can be improved
(Nolte and Hodgkiss, 1975). One sign of such a capability would
be a gradual emergence of direction sensitivity in E-E neurons as
the stimulus is repeated. This conjecture implies that, in an
unanesthetized bat, repeated, weak tone pulses from a given direction should facilitate some E-E neurons and inhibit the responses
of others. Neurons that correspond to erroneous direction hypotheses should display decreased responses that could easily be mis-
interpreted as habituation to the stimulus.

Menabe, et al., state that the E-E neurons "integrate or sometimes even multiply faint signals originating from both ears for
effective detection of a target." The ambiguity in this observation
is similar to the functional ambiguity of the coincidence counter
in Fig. 7. If the outputs of the ears are multiplied and then integrated, the detector functions as though it lacks knowledge of the
echo waveform. If the outputs are passed through narrowband filters
and summed, the detector functions as though the echo waveform is
known. Better detection performance is expected in the latter case,
and the assumption that the echo waveform is known applies especially well to CF echolocation in the mustache bat.

The interpretation of I-E neurons as direction estimates is consistent with difference beam localization of targets when the echo waveform is known. The fact that this estimate is made for targets at relatively close range may indicate a capability to track moths during escape maneuvers (Roeder, 1970; Altes and Anderson, this volume). I-E neurons can also have an interference suppression function. This function will be discussed below under the heading of adaptive binaural interference suppression.

A close connection between detection and estimation has already been observed in target classification models, and (17) also illustrates such a connection. The matched filter response $\underline{r}^T\underline{m}(\theta)$ can be used for computing maximum likelihood direction and amplitude estimates, and the mean of an updated version of $p(\theta)$ can also be used as an angle estimate (Scharf and Nolte, 1977). These concepts may be exploited in animal echolocation. For example, each E-E neuron has a best amplitude as well as a detection capability, indicating a possible dual role as detector and amplitude estimator.

In summary:

(i)   E-E neurons can be used for amplitude estimation as well as for detection. The E-E neurons that were investigated by Menabe, et al., probably correspond to a detector that has knowledge of the echo waveform, but little prior knowledge of target direction. An adaptive detector would utilize E-E neurons that become increasingly angle sensitive as the stimulus is repeated.

(ii)   I-E neurons probably correspond to a direction estimator for a known echo waveform. An I-E response can also be used as an error signal for an angle tracker, and for interference suppression (spatial filtering).

Another interesting neurophysiological result is the discovery of neurons, in the midbrain of the owl, that are excited by a sound stimulus from a given direction and inhibited by sound from slightly different directions (Knudsen and Konishi, 1978). This excitatory center, inhibitory surround organization has been observed in the optical cortex of mammals when the stimulus is a point of light that emanates from various directions (Hubel and Wiesel, 1968). In vision, this neural point-spread function has been interpreted as a sharpening device for accurate estimation of line and edge position. In Part IV, an interpretation in terms of principal components was suggested.

An excitatory center, inhibitory surround function can be obtained from a pressure difference sensor by spatial differentiation (Altes, 1978a). This observation suggests a mathematical relation between the midbrain neurons of the owl and the cortical I-E neurons

of the mustache bat. The differentiation operation is a spatial
high-pass filter that tends to accentuate angular inhomogeneities
in pressure.

The visual analogy suggests that valuable insights may be
obtained by thinking of animal sonar as an acoustic imaging system
that attempts to "see with sound." The point spread function of an
imaging system determines its resolution capability. An acoustic
counterpart of the point spread function is the range-angle ambiguity
function, which determines sonar position resolution in range and
direction.

d. The Range-Angle Ambiguity Function

The range-<u>angle</u> ambiguity function (Altes, 1978c, 1979) is a
generalization of the usual range-<u>Doppler</u> ambiguity function of
radar theory (Woodward, 1964). For an array of filters (or neurons)
that correspond to different range and angle hypotheses, a point
target will give rise to a characteristic pattern of filter (neuronal) responses as its range and angle are varied. This pattern
(which sometimes resembles an excitatory center with inhibitory
surround) is described by the range-angle ambiguity function.

When the volume and sidelobe level of the ambiguity function
are reduced, range and angle resolution improves. The volume of the
range-angle ambiguity function is inversely proportional to the number of array elements multiplied by the bandwidth of the signal.
This volume dependence suggests a trade-off between array size and
signal bandwidth. Given the limited array size of echolocating animals, the use of very wide bandwidths (in comparison to many existing man-made systems) can be understood in terms of an acoustic
imaging device.

The range-angle ambiguity function also leads to a generalization of the SIR formulation (6) to include angle-dependent clutter.

e. Adaptive Binaural Interference Suppression

Ideal binaural estimator/detector configurations have been
described. These configurations assume white, Gaussian background
noise that is directionally uniform (isotropic). If noise from a
particular direction is introduced, the ideal receiver spatially
"whitens" the data, i.e., the receiver attempts to obtain constant
noise power as a function of direction. In order to make the noise
isotropic, a null must be placed in the direction of the noise source.

Adaptive beam-forming methods for obtaining the desired null
have been described by many authors, e.g., Widrow, et al. (1967),

Frost (1972), Griffiths (1969), and Reed, Mallett, and Brennan
(1974). A biological counterpart of these beam-forming algorithms
is indicated by a phenomenon called binaural masking level difference (Green and Yost, 1975). Durlach (1963) has modelled this phenomenon with an equalization-cancellation device. In Durlach's
model (and in most adaptive beam formers) a null is obtained by
delaying the output of one ear, weighting it by an appropriate
factor, and subtracting it from the output of the other ear.

Neural analogs for this model of binaural interaction are the
I-E neurons of the mustache bat. I-E neurons may thus be implicated
in interference suppression as well as direction estimation. By
placing a null in the direction of an interfering signal, a sonar
effectively estimates the direction of the source of interference.

VII. PHASE SENSITIVITY IN ECHOLOCATION

a. Definition of Phase Sensitivity

Suppose that the Fourier transform of a signal u(t) is a
complex function

$$U(f) = A(f) \exp[j\phi(f)] . \tag{18}$$

A(f) is called the magnitude of U(f), and $\phi(f)$ is usually called the
phase. When we speak of phase insensitivity, however, we shall
mean that a signal $u_o(t)$ with Fourier transform

$$U_o(f) = A(f) \exp\{j[\phi(f) + \phi_o]\} \tag{19}$$

cannot be distinguished from u(t), for any constant phase parameter
ϕ_o. For example, if $\phi_0 = 180°$, then $u_0(t) = -u(t)$. Such a sign
reversal is apparently undetectable by a human listener. In this
chapter, the word "phase" refers to ϕ_0 in (19).

b. Why Measure Phase?

The phase shift that occurs when a sonar signal is reflected
from a target is a useful feature in both air and water.

In air, all targets have a higher acoustic impedance than the
propagation medium, and no information about target "hardness" is
obtained from a phase measurement. In Fig. 1, however, we saw
that a linear-period modulated (LPM) signal experiences a phase
shift when it is reflected from a moving target. Although relative
phase shifts between harmonics can be used for motion sensitivity,
high frequencies can be rapidly attenuated by absorption in air
(Griffin, 1971). The phase of the fundamental component of an

echo, measured with respect to the transmitted waveform, would therefore be useful for moving target indication at long range.

In water, some targets (like rocks) have higher acoustic impedance than the propagation medium, and echoes from such targets experience no phase reversal. Other targets, like air bladders of fish, have lower acoustic impedance than the medium, and echoes from these objects are phase shifted by 180°. In order to measure these impedance-induced phase shifts with linear-period modulated signals, Doppler-induced phase shifts should be avoided by using low chirp rates (only a few zero crossings). LPM dolphin clicks exhibit this property.

Finally, phase sensitivity implies that the detailed structure of a matched filter response can be measured, rather than the envelope of the response. This fine structure can yield superior range resolution (Simmons, 1979) and can lead to coherent summation of echoes for better long range detection (Vel'min, Titov, and Yurkevich, 1975).

c. Possible Mechanisms for Phase Measurement in Echolocation

Mammalian echolocation signals are typically ultrasonic, or at least above 6 kHz. For frequencies that are above 6 kHz, auditory nerve data ceases to contain phase information (Anderson, et al., 1971). There are at least two possible mechanisms to circumvent this problem.

The first method is to calculate an energy spectrum from time data that includes both the transmitted signal and the echo. Although the 200-300 μsec critical interval of Vel'min and Dubrovskiy (1976) is too short for this application, the integration times that were measured by C. S. Johnson (1968a) are ~0.10 second long. If the echo $u_o(t-\tau)$ from a point target or highlight is shifted by ϕ_0 radians with respect to the transmitted signal $u(t)$, the resulting energy spectrum is (Johnson and Titlebaum, 1976)

$$|U(f) + a\,U(f)\,\exp[j(\phi_o - 2\pi f\tau)]|^2 = (1+a^2)|U(f)|^2$$

$$+ 2a|U(f)|^2\cos(\phi_o - 2\pi f\tau) \qquad (20)$$

where a is an attenuation factor. Just as the energy spectrum is sensitive to the relative phase between two harmonics, it is sensitive to the relative phase between two pulses that are separated in time. The cosine term in (20) represents a ripple or amplitude modulation, and the locations of peaks and nulls of the energy spectrum depend upon ϕ_0.

The second method of phase measurement exploits low frequency noise components. If the data contains low frequency noise, then the phase of a noise component that is below 6 kHz can theoretically be measured. Since ϕ_0 is independent of frequency, this phase measurement theoretically allows the receiver to completely reconstruct the input data (signal plus noise) from the spectrogram of the data (Altes, 1978b,d). Subsequent matched filtering eliminates low frequency noise components, but the phase of the echo can, in theory, be extracted from the completely reconstructed data waveform.

d. Suggested Experiments

The above conjectures about phase sensitivity can be experimentally investigated. By using a transponder that re-transmits a phase-shifted version of an animal's signal back to the animal, one can test for phase sensitivity in echolocation. If phase cannot be measured by passive listening but only via active echolocation, then the first phase measurement method, using an energy spectrum with long integration time, may be a relevant mechanism.

The second method of phase measurement, which exploits low frequency noise components, suggests that past phase sensitivity measurements may have been "too clean." Elimination of low frequency noise destroys a receiver's theoretical capability to completely reconstruct an ultrasonic pulse from a spectrogram.

Another experimental approach is to measure sensitivity to the fine structure of a signal's autocorrelation function, since phase insensitivity implies that only the envelope of the autocorrelation function can be used. This approach has been investigated by Simmons (1979), and his range resolution data seem to indicate phase sensitivity in the FM bat Eptesicus fuscus. Sensitivity to fine structure is also indicated by the ability to coherently add pulses. Coherent addition gives a faster increase in SNR than is obtained by summing envelope detected pulses. Vel'min, Titov, and Yurkevich (1975) claim to have observed coherent pulse summation in dolphins, for a pulse repetition frequency in the range 1-6 pulses per second.

VIII. SUMMARY AND CONCLUSION

Animal echolocation has been interpreted in terms of engineering problems such as estimation, detection, classification, pattern recognition, array processing, interference suppression, and description of range extended, time-varying sonar targets. Adaptive solutions to these problems have been accentuated because such solutions are becoming more important for man-made systems and because biological systems sometimes display adaptive behavior. Adaptivity is

also important because ideal receiver configurations depend upon
what is known or unknown (e.g., the echo waveform and target location), and this knowledge can change with time.

The three most challenging and important tasks have been
(i) to decide what problem an animal is actually solving (e.g.,
what target features should be estimated), (ii) to decide what
constraints exist on the animal's solution to the problem (e.g.,
monaural phase deafness), and (iii) to make quantitative performance predictions so that a model can be checked against experimental data.

Some objections to modelling were given in the introduction,
but counter-examples have been found for most of them:

The feeling that mathematics may be unnecessary for an appreciation of biology is belied by the need for statistical communication theory in order to understand (i) the relation of biological
data to behavior, i.e., to detection, estimation, and classification
functions, and (ii) the transfer of information within the auditory system via neural encoding. An impressive example of non-
mathematical insight can be found in the functional interpretation
of E-E and I-E neurons by Menabe, Suga, and Ostwald (1978). Even
in this case, however, theoretical models help to fully specify the
detection problem that is being solved by the bat and point to the
possible existence of an adaptive detector as well as a device for
interference suppression.

A premature theoretical construction, based upon sketchy data,
is perhaps illustrated by the model that uses the geometrical theory
of diffraction, along with a receiver complexity constraint. Quantitative analysis and experimental performance of the model have
indicated a need for revision. The theory has thus been twisted
to fit facts, rather than vice versa. In the course of finding
fault with the theory, previously unavailable echo data had to be
gathered, and these data have served as building blocks for an
improved model. Tests of the model have also led to some important
concepts, such as the use of near-orthogonal basis functions for
echo analysis, and the difficulty of analyzing separate scattering
centers that have overlapping echoes.

Although a model may be restrictive in its over-simplification
of reality, the simultaneous consideration of several models can
significantly increase our understanding. By comparing Colburn's
model for binaural interaction (Fig. 7) with theoretical systems
for detection and estimation, we have spotlighted the importance
of the operation that is performed by the coincidence counter. By
applying pattern recognition principles to spectrograms, we may
have obtained insight into the function of excitatory center, inhibitory surround neurons in vision and sound localization. Each of

these concepts corresponds to one restricted viewpoint of reality. An analogous situation is found in the story of the blind men, each of whom perceives an elephant as a different object because of his own unique encounter with the animal (without the benefit of ultra-sonic spectacles). A concerted effort by an organized, communicative group of blind men should eventually solve the problem.

Beauty and elegance appear in some of the models, e.g., in the use of a scattering function and its relation to spectrograms. But no matter how elegant a theory is, we are still obliged to search for experiments to test its validity, since a theory is vacuous without such verification. We are thus unlikely to become overly attached to a model that is fundamentally erroneous.

"Our objective is to abstract patterns from Nature..., but many proposed patterns do not in fact correspond to the data. Thus all proposed patterns must be subjected to the sieve of critical analysis, and rigid skepticism without a search for patterns, are the antipodes of incomplete science. The effective pursuit of knowledge requires both functions." (Sagan, 1977).

IX. REFERENCES

Ackroyd, M. H., 1971, Short-time spectra and time-frequency energy distributions, J. Acoust. Soc. Amer., 50:1229.

Ahmed, N., and Rao, K. R., 1975, "Orthogonal Transforms for Digital Signal Processing", Springer-Verlag, Berlin.

Altes, R. A., 1971a, Methods of wideband signal design for radar and sonar systems, Fed. Clearinghouse N° AD 732-494.

Altes, R. A., 1971b, Suppression of radar clutter and multipath effects for wideband signals, IEEE Trans. on Inform. Theory, IT-17:344.

Altes, R. A., 1976a, Bionic image analysis using lines and edges, Math. Biosciences, 31:317.

Altes, R. A., 1976b, Sonar for generalized target description and its similarity to animal echolocation systems, J. Acoust. Soc. Amer., 59:97.

Altes, R. A., 1977, Estimation of sonar target transfer functions in the presence of clutter and noise, J. Acoust. Soc. Amer., 61:1371.

Altes, R. A., 1978a, Angle estimation and binaural processing in animal echolocation, J. Acoust. Soc. Amer., 63:155.

Altes, R. A., 1978b, "Further Development and New Concepts for Bionic Sonar. Vol. 2. Spectrogram Correlation", Report OC-R-78-A004-1, ORINCON Corporation, 3366 N. Torrey Pines Ct., La Jolla, Ca., Reprinted as NOSC TR-404, Naval Ocean Systems Center, San Diego, Ca.

Altes, R. A., 1978c, "Further Development and New Concepts for Bionic Sonar. Vol. 3. New Concepts and Experiments", see 1978b.

Altes, R. A., 1978d, Possible reconstruction of auditory signals
    by the central nervous system, J. Acoust. Soc. Amer., 64,
    Supp. 1, S137.

Altes, R. A., 1979a, Target position estimation in radar and sonar,
    and generalized ambiguity analysis for maximum likelihood
    parameter estimation, Proc. IEEE, 67.

Altes, R. A., 1979b, Utilization of spectrograms for detection and
    estimation, with applications to theories of hearing and
    animal echolocation. Part I. Spectrogram processing, sub-
    mitted for publication to J. Acoust. Soc. Amer.

Altes, R. A., and Anderson, G. M., Binaural estimation of cross-
    range velocity and optimum escape maneuvers by moths, this
    volume.

Altes, R. A., and Faust, J. W., 1978, "Further Development and New
    Concepts for Bionic Sonar. Vol. 1. Software Processors",
    see Altes, 1978b.

Altes, R. A., and Reese, W. D., 1975, Doppler-tolerant classification
    of distributed targets--a bionic sonar, IEEE Trans. on Aero-
    space and Electronic Systems, AES-11:708.

Altes, R. A., and Skinner, D. P., 1977, Sonar velocity resolution
    with a linear-period-modulated pulse, J. Acoust. Soc. Amer.,
    61:1019.

Altes, R. A., and Titlebaum, E. L., 1970, Bat signals as optimally
    Doppler tolerant waveforms, J. Acoust. Soc. Amer., 48:1014.

Altes, R. A., and Titlebaum, E. L., 1975, Graphical derivations
    of radar, sonar, and communication signals, IEEE Trans. on
    Aerospace and Electronic Systems, AES-11:38.

Anderson, D. J., Rose, J. E., Hind, J. E., and Brugge, J. F., 1971,
    Temporal position of discharges in single auditory nerve
    fibers within the cycle of a sine wave stimulus: frequency
    and intensity effects, J. Acoust. Soc. Amer., 49:1131.

Bangs, W. J., and Schultheiss, P. M., 1973, Space-time processing
    for optimal parameter estimation, in: "Signal Processing",
    Griffiths, stocklin, and Van Schooneveld, eds., Academic
    Press, London.

Bechtel, M. E., 1976, Short pulse target characteristics, in: "Atmos-
    pheric Effects on Radar Target Identification and Imaging",
    H. E. G. Jeske, ed., Reidel, Dordrect.

Bechtel, M. E., and Ross, R; A., 1966, Radar scattering analysis,
    CAL report N° ER/RIS-10, Cornell Aeronautical Laboratory,
    Buffalo, N.Y.

Bello, P.A., 1963, Characterization of randomly time-variant linear
    channels, IEEE Trans, on Comm. Sys., CS-11:360

Bel'kovich, V. M., and Dubrovskiy, N. A., 1976, "Sensory Bases of
    Cetacean Orientation", Chapter V. Echolocation, JPRS, L/7157.

de Boer, E., 1975, Synthetic whole-nerve action potentials for the
    cat, J. Acoust. Soc. Amer., 58:1030.

Brennan, L. E., and Reed, I. S., 1973, Theory of adaptive radar,
    IEEE Trans. Aerosp. and Electronic Systems, AES-9:237.

Bullock, T. H., Grinnell, A. D., Ikezono, E., Kamseda, K., Katsuki, Y., Nomoto, M., Sato, O., Suga, N., Yanagisawa, K., 1968, Electrophysiological studies of central auditory mechanisms in cetaceans, Zeitschrift für Vergleichende Physiologie, 59: 117.

Cahlander, D. A., 1966, Echolocation with wideband waveforms: bat sonar signals, Fed. Clearinghouse N° AD 605-322.

Capon, J., 1961, On the asymptotic efficiency of locally optimum detectors, IRE Trans. Inform. Theory, IT-7:67.

Chien, Y. T., and Fu, K. S., 1968, Selection and ordering of feature observations in a pattern recognition system, Inform. and Control, 12:395.

Colburn, H. S., 1973, Theory of binaural interaction based on auditory nerve data. I. General strategy and preliminary results on interaural discrimination, J. Acoust. Soc. Amer., 54:1458.

Colburn, H. S., 1977, Theory of binaural interaction based on auditory nerve data. II. Detection of tones in noise, J. Acoust. Soc. Amer., 61:525.

Colburn, H. S., and Latimer, J. S., 1978, Theory of binaural interaction based on auditory nerve data. III. Joint dependence on interaural time and amplitude differences, J. Acoust. Soc. Amer., 64:95.

Cook, C. E., and Bernfeld, M., 1967, "Radar Signals", Academic Press, New York.

Davenport, W. B., and Root, W. L., 1958, "Random Signals and Noise", McGraw-Hill, New York.

Davis, R. C., Brennan, L. E., and Reed, I. S., 1976, Angle estimation with adaptive arrays in external noise fields, IEEE Trans. on Aerospace and Electronic Systems, AES-12:179.

Decouvelaere, M., 1979, Signal design for matched filter detection in a reverberation-limited environment: application to cetacean echolocation signals, this volume.

De Long, D. F., and Hofstetter, E. M., 1969, The design of clutter-resistant radar waveforms with limited dynamic range, IEEE Trans. on Inform. Theory, IT-15:376.

Dirac, P. A. M., 1963, The evolution of the physicist's picture of nature, Scientific American, 208, N° 5:45.

Durlach, N. I., 1963, Equalization and cancellation theory of binaural masking-level differences, J. Acoust., Soc. Amer., 35:1206.

Dziedzic, A., and Alcuri, G., 1977, Reconnaissance acoustique des formes et caracteristiques des signaux sonars chez Tursiops truncatus, C. R. Acad. Sc. Paris, 285 D: 981.

Evans, E. F., 1977, Peripheral processing of complex sounds, in: "Recognition of Complex Acoustic Signals", T. H. Bullock, ed., Abakon Verlagsgesellschaft, Berlin.

Evans, W. E., 1973, Echolocation by marine dolphinids and one species of fresh-water dolphin, J. Acoust. Soc. Amer., 54:191.

Feinman, R., 1974, Structure of the proton, Science, 183:601.

Freedman, A., 1962, A mechanism of acoustic echo formation, Acustica, 12:10.

Frost, O. L., 1972, An algorithm for linearly constrained adaptive array processing, Proc. IEEE, 60:926.

Glaser, E. M., 1961, Signal detection by adaptive filters, IRE Trans. on Inform. Theory, IT-10:87.

Green, D. M., 1958, Detection of multiple component signals in noise, J. Acoust. Soc. Amer., 30:904.

Green, D. M., McKey, M. J., and Licklider, J. C. R., 1959, Detection of a pulsed sinusoid in noise as a function of frequency, J. Acoust. Soc. Amer., 31:1446.

Green, D. M., and Swets, J. A., 1966, "Signal Detection Theory and Psychophysics, Wiley",New York.

Green, D. M., and Yost, W. A., 1975, Binaural analysis, in: "Handbook of Sensory Physiology",Vol. 5, Part 2, Springer-Verlag, Berlin.

Greville, T. N. E., 1969, "Theory and Applications of Spline Functions", Academic Press, New York.

Griffin, D. R., 1958, "Listening in the Dark", Yale University Press, New Haven, Conn.

Griffin, D. R., 1971, The importance of atmospheric attenuation for the echolocation of bats (Chiroptera), Anim. Behav., 19:55.

Griffiths, L. J., 1969, A simple adaptive algorithm for real-time processing in antenna arrays, Proc. IEEE, 57:1696.

Hahn, W. R., and Tretter, S. A., 1973, Optimum processing for delay-vector estimation in passive signal arrays, IEEE Trans. Inform. Theory, IT-19:608.

Hahn, W. R., 1975, Optimum signal processing for passive sonar range and bearing estimation, J. Acoust. Soc. Amer., 58:201.

Harger, R. O., 1970, "Synthetic Aperture Radar Systems", Academic Press, New York.

Hauske, G., Wolf, W., and Lupp, U., 1976, Matched filters in human vision, Biol. Cybernetics, 22:181.

Hodgkiss, W. S., 1978, Detection of LPM signals with estimation of their velocity and time of arrival, J. Acoust. Soc. Amer., 64:177.

Hubel, D. H., and Wiesel, T. N., 1968, Receptive fields and functional architecture of monkey striate cortex, J. Physiol., 195: 215.

Jeffress, L. A., 1948, A place theory of sound localization, J. Comp. Physiol. Psychol., 41:35.

Johnson, C. S., 1968a, Relation between absolute threshold and duration-of-tone pulses in the bottlenosed porpoise, J. Acoust. Soc. Amer., 43:757.

Johnson, C. S., 1968b, Masked tonal thresholds in the bottlenosed porpoise, J. Acoust. Soc. Amer., 44:965.

Johnson, R. A., 1972, "Energy spectrum analysis as a processing mechanism for echolocation", Ph. D. Diss., University of Rochester, New York.

Johnson, R. A., and Titlebaum, E. L., 1976, Energy spectrum analysis: a model of echolocation processing, J. Acoust. Soc., Amer., 60:484.

Kassam, S. A., and Thomas, J. B., 1976, Dead-zone limiters: an application of conditional tests in nonparametric detection, J. Acoust. Soc. Amer., 60:857.

Kassam, S. A., and Thomas, J. B., 1977, Improved nonparametric coincidence detectors, J. Franklin Inst., 303:75.

Knudsen, E. E., and Konishi, M., 1978, Space and frequency are represented separately in the auditory midbrain of the owl, J. Neurophysiol., 41:870.

Koestler, A., 1964, "The Act of Creation", Dell, New York.

Kotelenko, L. M., and Radionova, E. A., 1975, On the phase sensitivity of neurons in the cat's auditory system, J. Acoust. Soc. Amer., 57:979.

Kouyoumjian, R. G., 1965, Asymptotic high-frequency methods, Proc. IEEE, 53:864.

Kroszczynski, J. J., 1969, Pulse compression by means of linear- period modulation, Proc. IEEE, 57:1260.

Lee, S. D., and Uhran, J. J., 1973, Optimum signal and filter design in underwater acoustic echo ranging systems, IEEE Trans. on Aerospace and Electronic Systems, AES-9:701.

Licklider, J. C. R., 1959, Three auditory theories, in: "Psychology: A Study of a Science", S. Koch, ed., McGraw-Hill, New York.

Livshits, M. S., 1974, Some properties of the dolphin hydrolocator from the viewpoint of a correlation hypothesis, Biofizika, 19:916, JPRS 64329.

Makhoul, J., 1975, Linear prediction: a tutorial review, Proc. IEEE, 63:561.

Menabe, T., Suga, N., and Ostwald, J., 1978, Aural representation in the Doppler-shifted-CF processing area of the auditory cortex of the mustache bat, Science, 200:339.

Middleton, D., 1966, Canonically optimum threshold detection, IEEE Trans. on Inform. Theory, IT-12:230.

Miller, E. K., Deadrick, F. J., Hudson, H. G., Poggio, A. J., and Landt, J. A., 1975, Radar target classification using temporal mode analysis, report UCRL-51825, Lawrence Livermore Laboratory, University of California, Livermore Ca.

Nilsson, H. G., 1978, A comparison of models for sharpening of the frequency selectivity in the cochlea, Biol. Cybernetics, 28: 177.

Nolte, L. W., and Kodgkiss, W. S., 1975, Directivity or adaptivity?, EASCON '75, (35-A)-(35-H).

Patterson, J. H., and Green, D. M., 1970, Discrimination of transient signals having identical energy spectra, J. Acoust. Soc. Amer., 48:894.

Picinbono, B., 1978, Adaptive signal processing for detection and communication, in: "Communication Systems and Random Process Theory", Skwirzynski, ed., Sijthoff and Hoordhoff, Alphen aan den Rijn.

Reed, I. S., Mallett, J. D., and Brennan, L. E., 1974, Rapid con-
    vergence rate in adaptive arrays, IEEE Trans. on Aerospace
    and Electronic Systems, AES-10:853.

Rihaczek, A. W., 1968, Signal energy distribution in time and fre-
    quency, IEEE Trans. on Inform. Theory, IT-14:369.

Rihaczek, A. W., 1969, "Principles of High-Resolution Radar",
    McGraw-Hill, New York.

Roberts, R. A., 1965, On the detection of a signal known except
    for phase, IEEE Trans. Inform. Theory, IT-11:76.

Roeder, K. D., 1970, Episodes in insect brains, Amer. Scientist,
    58:378.

Rummler, W. D., 1966, Clutter suppresssion by complex weighting of
    coherent pulse trains, IEEE Trans. on Aerospace and Electronic
    Systems, AES-2;689.

Russell, B., 1927, "An Outline of Philosophy", Allen and Unwin,
    London.

Sagan, C., 1977, "The Dragons of Eden", Ballantine Books, New York.

Sayers, B., McA., and Cherry, E. C., 1957, Mechanism of binaural
    fusion in the hearing of speech, J. Acoust. Soc. Amer., 29:
    973.

Scharf, L. L., and Nolte, L. W., 1977, Likelihood ratios for sequen-
    tial hypothesis testing on Markov sequences, IEEE Trans. on
    Inform. Theory, IT-23:101.

Schnitzler, H.-U., 1973, Control of Doppler shift compensation in
    the greater horseshoe bat Rhinolophus ferrumequinum, J. Comp.
    Physiol., 82:79.

Siebert, W. M., 1968, Stimulus transformations in the peripheral
    auditory system, in: "Recognizing Patterns", Kolers and Eden,
    eds., MIT Press, Cambridge, Mass.

Siebert, W. M., 1970, Frequency discrimination in the auditory sys-
    tem: place or periodicity mechanisms?, Proc. IEEE, 58:723.

Simmons, J. A., 1979a, Perception of echo phase information in bat
    sonar, in press.

Simmons, J. A., 1979b, Processing of sonar echoes by bats, this
    volume.

Skinner, D. P., Altes, R. A., and Jones, J. D., 1977, Broadband
    target classification using a bionic sonar, J. Acoust. Soc.
    Amer., 62:1239.

Sparks, D. W., 1976, Temporal recognition masking--or interference?,
    J. Acoust. Soc. Amer., 60:1347.

Spilker, J. J., and Magill, D. T., 1961, The delay-lock discrimina-
    tor--an optimum tracking device, Proc. IRE, 49:1403.

Stern, R. M., and Colburn, H. S., 1978, Theory of binaural inter-
    action based on auditory nerve data. IV. A model for sub-
    jective lateral position, J. Acoust. Soc. Amer., 64:127.

Stutt, C. A., and Spafford, L. J., 1968, A "best" mismatched filter
    response for radar clutter discrimination, IEEE Trans. on
    Inform. Theory, IT-14:280.

Suga, N., 1972, Analysis of information-bearing elements in complex
        sounds by auditory neurons of bats, Audiology, 11:58.
Turin, G. L., 1957, On the estimation in the presence of noise of
        the impulse response of a random, linear filter, IRE Trans.
        on Inform. Theory, IT-3:5.
Van Trees, H. L., 1968, "Detection, Estimation, and Modulation
        Theory, Part I", Wiley, New York.
Van Trees, H. L., 1971, "Detection, Estimation, and Modulation
        Theory, Part III", Wiley, New York.
Vakman, D. E., 1968, "Sophisticated Signals and the Uncertainty
        Principle in Radar", Springer-Verlag, New York.
Vel'min, V. A., and Dubrovskiy, N. A., 1976, The critical interval
        of active hearing in dolphins, Soc. Phys. Acoust., 22:351.
Vel'min, V. A., Titov, A. A., and Yurkevich, L. I., 1975, Temporal
        pulse summation in bottlenosed dolphins, in: "Kiev Morskiye
        Mlekopitayushchiye", Agarkov, ed.
Widrow, B., 1962, Generalization and information storage in net-
        works of adaline "neurons", in: "Self-Organizing Systems--
        1962", Yovits, Jacobi, and Goldstein, eds., Spartan Books,
        Washington, D. C.
Widrow, B., Mantey, P. G., Griffiths, L. J., and Goode, B. B., 1967,
        Adaptive antenna systems, Proc. IEEE, 55:2143.
Wolff, S. S., Thomas, J. B., and Williams, T. R., 1962, The polarity-
        coincidence correlator: a non-parametric detection device,
        IRE Trans. on Inform. Theory, IT-8:5.
Woodward, P. M., 1964, "Probability and Information Theory, With
        Applications to Radar", Pergamon Press, Oxford.

ENERGY SPECTRUM ANALYSIS IN ECHOLOCATION

Richard A. Johnson

Naval Ocean Systems Center

San Diego, California   92152

INTRODUCTION

There are many approaches to the problem of understanding echolocation. Classically, observation and behavioral experimentation have been used to define echolocating animals' abilities, while more recently emphasis has shifted to neurophysiological investigations, particularly with certain bat species. However, details of the acoustic processing mechanisms involved remain obscure, although inferences can be made concerning the efficiency of extracting information from echoes.

Several theories have been proposed to explain the neurophysiological processing necessary for acoustic orientation. In order to infer the distance to objects for example, it is necessary to know the round trip travel time Δ associated with the acoustic propagation from the animal to the object and back. The distance d to the object is

$$d = (\Delta \cdot Vp)/2$$

where Vp is the velocity of acoustic propagation in the medium. Most simply, it has been theorized that Δ is directly measured neurophysiologically, roughly equivalent to neural potential spikes stimulated by acoustic pulses. However, many uncertainties such as noise, weak echo strength, and echo distortions make the direct determination of Δ by neural networks a rough estimate, at best. Slightly more complex theories, such as Kay's frequency modulation range-finder (Kay, 1967) or Pye's beat notes (Pye, 1967), are very restrictive in that they do not allow for the slight inaccuracies in the generation of signals which one finds

in biological systems. Nor do they account for non-FM signals.
However, at the same time, Pye recognized the utility of wide-
bandwidth pulses, spectral interference patterns, and the minimal
effect of Doppler on FM pulses. Consideration of the components
and functioning of man-made radar and sonar has led to more recent
theoretical emphasis on correlation and matched filter processing.
Encouragement was provided by signal derivation studies (Altes &
Titlebaum, 1970; Altes & Reese, 1975) and animal performance tests
(Simmons, 1970, 1971a, 1971b, 1973; Simmons, Howell & Suga, 1975;
Murchison, this volume).

The main concerns of this paper are object detection, object
distance estimation, and object identification and how they may be
accomplished via energy spectrum analysis as an alternative to
correlation processing in the time domain sense. With the additional
cue of object direction, the task of acoustic orientation can be
satisfactorily explained.

INTERFERENCE PATTERNS IN SPECTRAL ANALYSIS

Interference patterns are well-known wave phenomena. Examples
include Newton's rings (optics), two coherent sources (optics,
ripple tanks) and beat notes (acoustics). While the causes may
vary, the result is the same: alternating constructive and destructive interference phenomenon with applications to echolocation.

Laboratory analysis of echolocation sounds can involve both
time domain techniques (oscilloscopes, period meters) and frequency
domain techniques (sonagraphs, FET's, spectral analyzers). The
latter represent implementations of Fourier analysis with considerations for integration time, frequency resolution, dynamic range,
etc. Also, usually only a representation of the magnitude squared
of the Fourier transform known as the energy spectrum is retained
from this analysis and often the time history of the energy spectrum
is displayed as a sonagram or waterfall spectrogram. The energy
spectrum can display interference patterns under certain conditions
(Johnson and Titlebaum, 1976):

Let $u(t)$ be the time domain representation of a finite
duration signal, and let $U(f)$ be the Fourier transform of $u(t)$ so
that $|U(f)|^2$ is the energy spectrum of $u(t)$.

Now the signal plus a time delayed replica of the signal is
represented by $u(t) + u(t-\Delta)$. Using well-known properties of
Fourier transforms, one finds that the transform of this signal
pair is represented by

$$|U(f)|^2 \ [1 + \exp \ (-2\pi f \Delta)]$$

and the energy spectrum is represented by

$$|U(f)|^2 [2 + 2 \cos (2\pi f \Delta)].$$

This indicates that the original energy spectrum is simply modified
by the multiplying term $2 + 2 \cos (2\pi f\Delta)$ which is an interference
pattern. The number of cycles present is a function of how wide
the original spectrum $|U(f)|^2$, is and the time delay, Δ. An
example illustrating this effect is depicted in figure 1. More
generally, this interference pattern, or rippled spectrum (Yost,
et al, 1978), can have a lesser amplitude as a function of the
ratio of the two pulse amplitudes and the coherence of the two
pulses. Also, several ripple patterns can be present simultaneously
as a function of the number of pulses present (Johnson and Titlebaum,
1976). As depicted in figure 1, the delay parameter, Δ, is represented by the reciprocal of the distance between peaks of the
ripple.

ECHOLOCATION WITH WIDEBAND PULSES

     This spectral interference phenomenon has application as a
model of echolocation processing. An animal's outgoing pulse
corresponds to the first signal, while the returning echo (or
echoes) corresponds to the delayed signal (or signals). A spectrum
analysis processor can detect objects (ripple), determine the
distance to the object (a function of the distance between ripple
peaks) and possibly identify the echo source (discussed in more
detail in a later section). All of these analyses will be facilitated
if the energy spectrum of the outgoing pulse is very broad (large
bandwidth). Signal theory has shown that two factors can contribute
to large signal bandwidth (Johnson and Titlebaum, 1976). The
first factor is signal duration; that is, the briefer a signal is,
the broader the bandwidth. The second factor is a large change in
the instantaneous frequency over the signal duration; that is,
frequency modulation or FM. Also, note that harmoic components by
themselves do not necessarily contribute to large bandwidths, but
when combined with either of the above two factors, can significantly
broaden the energy spectrum. Apparently without exception, all
echolocating animals are capable of using signals that meet this
large bandwidth criterion and for many, broadband signals are used
exclusively.
     Spectral analysis of broadband signals as an explanation for
echolocation processing has several advantages over alternative
hypotheses. First of all, the process is correlation-equivalent
(Johnson and Titlebaum, 1976). As such, signal derivations and
concepts relating to correlation processing or matched filtering
continue to be applicable. (Loosely speaking, one might envision
the interference pattern as a type of correlation.) For example,
Doppler-tolerant bat signals will have well-behaved ripple peak

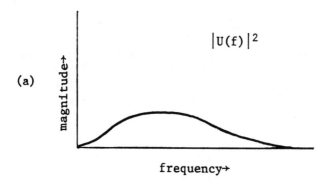

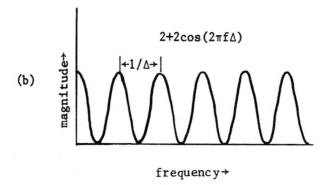

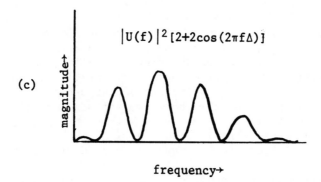

Figure 1.   (a) Signal spectrum, (b) Interference pattern,
            (c) Interference spectrum of signal pair

separations as well as well-behaved ambiguity functions (Johnson, 1972). However, the improvements over matched filtering or correlation as normally implemented in signal processing schemes include flexibility in the signal generation mechanism and not requiring an exact stored time replica of the transmitted signal as normally implemented in correlators.

A second benefit stems from the animal hearing its own outgoing pulse as well as the echoes for certain tested bats (Pollak et al., 1972) and since exclusion of the outgoing pulse from the auditory system would be anatomically difficult, if not impossible, in any animal, it should not be surprising to find that the exact nature of each pulse would be useful for echolocation. The extent to which the level of the outgoing pulse is controlled by the auditory system (Henson, 1970; Pollak and Henson, 1973) supports spectrum analysis in that it is the relative levels of the pulse and echoes that determine the level of the spectral ripple, and benefits will accrue to a processor that can more closely match the level of the outgoing pulse to the returning echoes.

A third advantage is that neurons are only called upon to measure the spectral peak spacings, so that there would appear to be no frequency limit beyond the physical responses of the middle ear and cochlea (Pye, 1967).

Finally, as a unifying theory for echolocation, it suggests that the general mammalian auditory system possesses the rudimentary ability to echolocate, and it should not be surprising to find that the echolocating animals have refined the processing on particular types of pulses contained in a large class of signals suitable for the basic mechanism. The pathways for evolutionary development are represented by the broad variety of signals, capabilities, and uses established for echolocation. This does not mean that other types of processing are excluded from use by echolocating animals. For example, the constant frequency portions of certain bat pulses must be processed in another manner. Also, long range echolocation where round-trip travel times exceed perhaps 20 msec is difficult to include in this theory since the interference patterns become difficult to resolve and integration times are extremely long for the spectrum analysis of the pulse and echoes.

It is important at this point to note that faulty interpretation of correlation theory can lead to troublesome interpretations of echolocation behavior. The example which often causes problems is the so-called range-Doppler coupling in echolocating signals. This is manifested as a "tilt" in the ambiguity function diagram ridge. The implicit assumption of constant relative velocity between the animal and object allows a more plausible interpretation. Instead of arbitrarily choosing a time origin for an echolocation signal, choose an origin for which there is no range-Doppler coupling. This can always be done for the constant relative velocity assumption. This "uncoupled" origin will usually be apart from the signal, and it is the round-trip travel time of this origin which is measured. Hence, the true position of the object is indicated for a time different from the actual time during which the signal is reflected. Under the constant velocity assumption, this will make no difference. It has been shown (Johnson and Titlebaum, 1972) that for the linear period signals, this "uncoupled" origin is maximally close to the signal itself. The rate of frequency change will further improve this factor. Thus, instead of measuring range-Doppler coupling, it is more appropriate in echolocation to measure the time estimation error, which is the distance from the "uncoupled" origin of a signal to the signal itself. Note that for frequency downsweeps, this origin always preceeds the signal. The opposite is true for frequency upsweeps. Were an echolocating animal to use upsweeps, it would be predicting (under the constant relative velocity assumption) where an object will be. Since this is not generally the case, the indication of where the object was (within a very few milliseconds) would be the case for a correlation-type processor.

PITCH OF RIPPLED SPECTRA

     Evidence for this spectrum analysis of rippled spectra is provided by the psychoacoustic phenomenon of repetition pitch or time-separation pitch (TSP). While it has been associated with echolocation previously (Nordmark, 1960; Pye, 1967), only more recently has the link to a correlation-type processing been emphasized. In fact, the early conjecture that animals heard a "time-difference tone", possibly in a frequency range below their hearing range, was cause for much initial skepticism. However, psychoacoustically the pitch is not maskable in the usual sense and therefore must be arising centrally. Also, it is not necessary to conjecture that echolocating animals hear a pitch, only that they process rippled spectra in a similar manner and ultimately perceive information about their environment. The pitch effect does present useful analogies when applied to echolocation.

     Wilson (1967) summarized the psychoacoustics of delayed stimuli in the first animal sonar symposium, and others have elaborated on specific points since then (Bilsen and Ritsma, 1969; Ritsma, 1967; Wightman, 1973). In a manner similar to Yost et al (1978), I have also investigated the discriminability of TSP. The motive for doing so was to get an analogy to range discriminability in echolocation and to probe the processing mechanisms. Briefly,

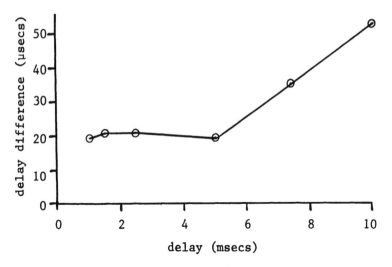

Figure 2. Performance of human subjects for time discrimination
        experiment.

the results (Johnson, 1977 and unpublished) were as follows:

For random pulse-pairs occuring at about 30 per second and
for delays from 10 to 1 msec between pulse pairs, the pitch resolution
varied from 0.5 Hz to 19 Hz respectively, which is considerably
better than that found by Yost et al. The improvement may be
attributed to the exact stimulus used and the technique for obtaining
the threshold. Also, in a manner analogous to the range discrimination éxperiments in echolocation, the human subjects were allowed
to make A-B comparisons repeatedly as they saw fit until they were
ready to indicate their choice. All subjects felt that the discrimination thresholds reported by Yost et al were trivally easy to
discriminate by this method. For comparison with animal experiments,
time discrimination is plotted as a function of time delay in
Figure 2.

As I have pointed out elsewhere (Johnson, 1977), the comparison
to similar plots for echolocating animals is startling (see also
Murchison, this volume). My subjects, under almost ideal conditions,
could perform nearly as well as a dolphin and considerably better
than bats. The time resolution of human subjects generally increase
as a function of time, thus resembling the dolphin resolution
function. It is not known why the bat data do not indicate similar
function.

In an effort to probe the processing mechanism, pulses which
exhibit the Doppler-tolerant behavior (Figure 3) were developed.
The stimuli consisted of repetitions of either identical pulse
pairs or pulse pairs where the second pulse of the pair was a
Doppler-compressed version of the first pulse. The rather large

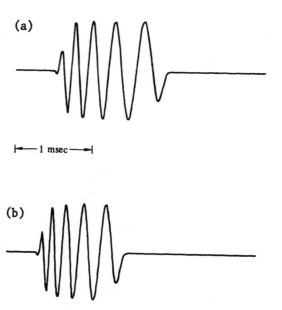

Figure 3. (a) Doppler tolerant pulse, (b) Non-normalized,
Doppler Compressed version of the pulse in (a)
with exactly one cycle shifted over

amount of compression was chosen so that the cross-correlation
function between the pulses was maximally like the autocorrelation
function of the uncompressed pulse (Fig. 4). The time delays were
chosen so that correlation processing would yield identical delay
estimates for the two cases. The result was that the pitch of the
two random pulse-pair trains was not identical. For a 2.5 msec
delay between pairs of identical pulses, the pitch did not become
the same for the non-identical pairs until the delay was increased
to 2.9 msecs. Several conclusions can be drawn from this. First,
time-domain correlation processing in the usual sense is unlikely
to be involved, given the specially constructed cross-correlation
function (Fig. 4). Second, other timing cues such as onset or
envelope characteristics do not correspond to the interval predicted.
Finally, since the corresponding spectra are different for the
different delay (Fig. 5), it is not obvious that a spectral processing
mechanism would produce the same pitch. Clearly not enough is
known about this processing to predict what pitch will be produced.

Perhaps the most valuable aspect of TSP experiments is experi-
encing this perception. There is no single written treatise that
best describes this complex auditory perception. In view of its
continuing association with echolocation processing, the value of
perceiving and even experimenting with TSP for oneself cannot be
overemphasized.

OBJECT IDENTIFICATION

As indicated in section III, it is also possible to explain
how spectrum analysis identifies objects. Before commenting on
how interference spectra can aid in this task, it is necessary to
review concepts in object identification, particularly as they
have been investigated for the U. S. Navy Bionic Sonar program.
The summary which follows is principally based on the work of
Chestnut and Landsman (1977) from that program.

As compared to conventional man-made sonars, a sonar identification system must be wideband. The frequency band of some dolphin
signals is 30 to 150 kHz, some bat signals cover 35 to 100 kHz,
and as a comparison, human speech perception uses the band from
200 to 3000 Hz. Conventional sonars use a band of less than 1
kHz. A basic postulate is that in order to perform target identification, the system must analyze the frequency response of the
target; hence, the system must have a wide frequency range.

Another postulate is that an identification system should
have good resolution of frequency components. Animals and humans
have good frequency resolution (Thompson and Herman, 1975), and
this seems intuitively to be a useful characteristic for recognition.

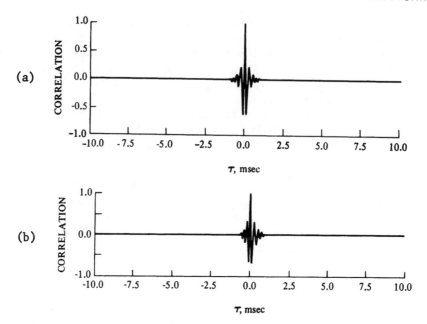

Figure 4. (a) Autocorrelation function for signal in Fig. 3a,
           (b) Crosscorrelation for signals of Figs. 3a and 3b.

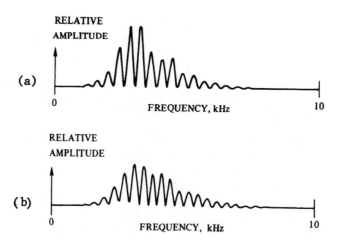

Figure 5. (a) Interference spectrum corresponding to uncom-
           pressed echo with 2.5 msec delay, (b) Interference
           spectrum corresponding to compressed echo with 2.9
           msec delay

Two types of features are used here to characterize the transfer functions of objects: energy in a bank of filters, and a mathematical model (the all-pole model) of the spectrum. While it may be impractical to store an accurate representation of a transfer function of an object, a sampled version may be adequate for object identification.

Suppose the frequency band of interest (f_a, f_b) is divided into subintervals defined by frequencies f_1, f_2, ..., f_n. The energy contained in the subinterval (f_{i-1}, f_i) is defined by

$$E_i = \int_{f_{i-1}}^{f_i} \left| \frac{S(f)}{U(f)} \right|^2 df$$

where $S(f)$ is the Fourier transform of the echo data and $U(f)$ is the Fourier transform of the signal. The values E_i, $= 1, 2, \ldots$, N are a sampled version of the energy distribution in (f_o, f_n).

The subintervals here are chosen in two ways: uniform and constant Q. The uniform intervals are defined by

$$f_o = f_a$$
$$f_i = f_{i-1} + \frac{f_b - f_a}{N}, \quad c = 1, 2, \ldots, N.$$

"Constant Q" means that the ratio of the bandwidths to center frequencies is a constant. In this case, the frequency intervals are defined by

$$f_o = f_a$$
$$f_i = f_{i-1} \cdot \frac{2Q+1}{2Q-1}, \quad = 1, 2, \ldots, N$$

where Q is the value of the ratio of center frequency to bandwidth for each subinterval. The value of Q is given in terms of f_a, f_b, and N by

$$Q = \frac{1}{2} \cdot \frac{(f_b/f_a)^{1/N} + 1}{(f_b/f_a)^{1/N} - 1}$$

There are three parameters which can be varied in this presentation of the transfer function: the frequency range (f_a, f_b), the

number of subintervals N, and the method of subdividing (constant
Q versus uniform).

Spectral modeling connotes the fitting of a mathematical
model to an experimentally derived spectrum. The spectrum is
represented, not by a set of discrete values of the spectrum or
its integral, but by a small number of parameters, such as the
poles and/or zeroes of a rational function, or the coefficients of
a polynomial. The model is defined as a continuous function of
frequency throughout the domain of the function. The advantages
of spectral modeling are that the amount of information required
to be stored may be smaller, and template matching is easier. The
information stored in the filter bank is highly redundant, because
it represents samples of a continuous function. Template matching
is easier because the comparison of model parameters replaces the
comparison of the actual spectra.

The all-pole model was chosen for the transfer functions of
objects. That is, the energy spectrum /S(f)/ of the data, when
divided by the energy spectrum U(f) of the transmitted signal, is
modeled by a function of the form

$$P(f) = \frac{G^2}{\left| 1 + \sum_{K=1}^{P} a_k e^{-2\pi jkf\Delta t} \right|^2}$$

where a_k, k = 1, ..., p and G are the parameters of the model;
Wt is the sampling period of the time data. The model is an all-
pole model because, in terms of the variable

$$z = e^{2\pi jk\Delta t}$$

is a rational function with poles and no zeroes. The coefficients
a_k of the model are obtained by the methods of linear prediction
and are called linear prediction coefficients (LPC), or reflection
coefficients.

The model is chosen because of its success in human speech
applications, and because it performs well with respect to spectral
maxima, perhaps caused by natural acoustic resonances of objects.
Also, the coefficients which are used to characterize the all-pole
model have the desirable properties of stability and independence;
that is, as the order of the model is increased, the lower order
coefficients do not change appreciably.

An all-pole model appears to be very good for delineating the
narrow spikes which are usually observed in the energy spectra of
echoes of common objects. The spikes arise from interference and

from natural object resonances. Other models require many more
coefficients to get equivalent approximations to these spectra.

Either type of feature extraction procedure results in a set
of numbers called a feature vector representing the transfer
function.

To evaluate these identifcation procedures, data were collected
from a set of objects, 13 cylinders of varying size and composition,
2 spheres and one bionic object. The signal used was a so-called
"bionic" signal introduced by Altes and Reese (1975) with parameters
chosen so that the center frequency was at 23 kHz, with a broad
bandwidth and short time duration (Fig. 6). Factors considered
important for this signal have no significance here, since the
effect of the specific signal is removed to obtain the transfer
function.

Both object identification techniques, one using energy in a
bank of filters, the other using the coefficients calculated from
the all-pole model of the transfer function, have been evaluated.
The performance criterion is the probability of misclassification
as a function of signal-to-noise ratio.

To establish the archetypal feature vectors for the set of
objects, a subset (called the design set) of the recorded echoes
was averaged, and the feature vector was calculated. This was
done for a variety of parameter settings for the two methods. In
each case, 4 echoes were used for this determination. The remaining
echoes (called the test set) were used to test the identification
method. Each echo not included in the design set was read, noise
was added, the feature vector for this noisy echo was calculated,
and the result was compared to the collection of archetypes. The
echo was associated with that object whose archetypal feature
vector was closest, in some metric, to the feature vector for that
object. If the minimum distance was too large, the sample was
rejected as not corresponding to any of the archetypal objects.
The probability of misclassification was estimated to be the
fraction of echoes that were associated with an incorrect object.
The probability of rejection was estimated to be that fraction of
the test echoes that were rejected. Of three metrics chosen, only
the data for the conventional Euclidean distance will be represented.
Other measures attempted were more sensitive to the added noise.
Parameters which were varied in the analysis were the frequency
band (15 to 45 kHz and 25 to 55 kHz) and the number of features
(5, 10, 15, and 30).

First, however, consider the results when no noise was added
to the test echoes (table 1). This table shows the fraction of
test echoes that were correctly identified using the various
methods indicated.

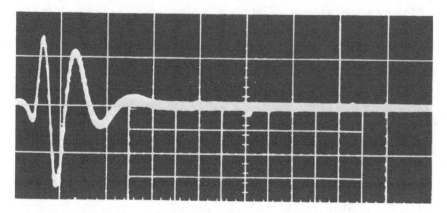

Figure 6.   Bionic Signal used for identification study.

The results show that good accuracy can be obtained with as few as 5 coefficients, and when 15 coefficients are used, some of the methods yield 100% correct identification. Energy detection achieves 100% when 15 or 30 filters are used, but this method degrades significantly when a smaller number of filters is used. The method based on the all-pole model is 100% accurate when 15 coefficients are used. The method based on the all-pole model is 100% accurate when 15 coefficient are used, and 99.4% correct when only 5 coefficients are used.

It is not advisable to make decisions about the relative merits of the various methods when the test is based upon data for which the signal-to-noise ratio is very high. A different picture is presented when noise is artificially created and added to the test echoes. The results of tests with noisy data are presented in figures 7 to 12. These figures show the probability of misclassification (dotted lines) and added to that the probability of rejection (solid lines) as a function of signal-to-noise ratio (SNR). If the signal power P after deconvolution is given by

$$P = \int_{f_1}^{f_2} \left| \frac{S(f)}{U(f)} \right|^2 \, df$$

and the expected noise power E(No) is given by

$$E\{No\} = s^2 \int_{f_1}^{f_2} \frac{1}{\left| U(f) \right|^2} \, df$$

Table I. The Fraction of Test Echoes which were Correctly
          Identified Using Various Methods, as a Function
          of the Number of Parameters Used

	5	10	15	30
Reflection coefficients, 15-45 kHz	0.994	0.989	1.00	
Reflection coefficients, 25-55 kHz	0.977	0.983	0.994	
Bank of uniform filters, 15-45 kHz	0.960	0.977	1.00	1.00
Bank of uniform filters, 25-55 kHz	0.932	0.994	1.00	1.00
Bank of constant Q filters, 15-45 kHz	0.920	0.983	1.00	1.00
Bank of constant Q filters, 25-55 kHz	0.909	1.00	0.994	1.00

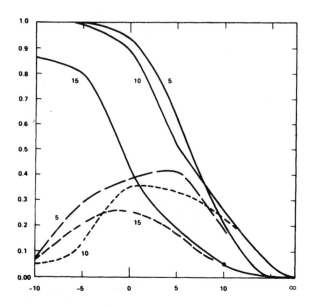

Figure 7.   Probability of Rejection and Misclassification as a
            Function of Signal-to-Noise Ratio for the All-Pole
            Model on 15-45 kilohertz Band Using the Euclidean
            Distance Measure

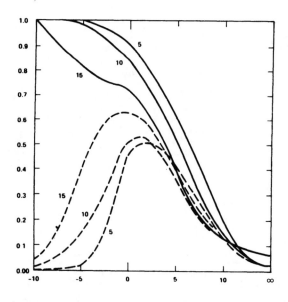

Figure 8.   Probability of Rejection and Misclassification as a
            Function of Signal-to-Noise Ratio for the All-Pole
            Model on 25-55 kilohertz Using the Euclidean Distance
            Measure

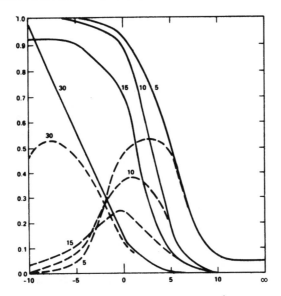

Figure 9. Probability of Rejection and Misclassification as a
           Function of Signal-to-Noise Ratio for a Bank of Constant
           Q Filters on the 15-45 kilohertz Band

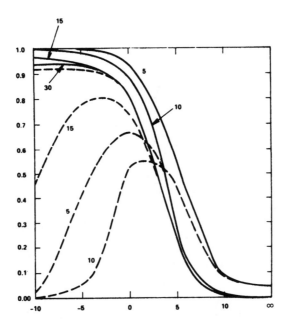

Figure 10. Probability of Rejection and Misclassification as a
            Function of Signal-to-Noise Ratio for a Bank of Constant
            Q Filters on the 25-55 kilohertz Band

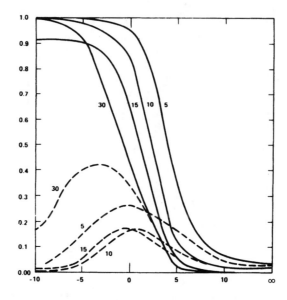

Figure 11. Probability of Rejection and Misclassification as a
Function of Signal-to-Noise Ratio for a Bank of Filters
of Uniform Width on the 15-45 kilohertz Band

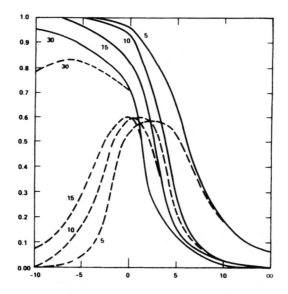

Figure 12. Probability of Rejection and Misclassification as a
Function of Signal-to-Noise Ratio for a Bank of Filters
of Uniform Width on the 25-55 kilohertz Band

where σ^2 is the expected value of the square of the magnitude of the Fourier transform of the noise, $N(f)$, then the SNR is defined to be

$$SNR = 10 \log_{10} (P/E\{No\})$$

Observe that as the SNR decreases, the probability of misclassification decreases and the probability of rejection tends to 1. This is to be expected because as the noise increases, the distance from the feature vector of an echo to every archetype increases. The threshold for rejection was chosen by observing the set of distances in the high SNR cases.

When noise is added to the echoes, the method which uses the energy in a bank of filters is superior to the method which uses the all-pole model. The best results are obtained using 30 filters on the interval from 15 to 45 kHz for these objects. There are only slight differences between the uniformly spaced filter bank and the constant Q filter bank.

The reason for the superiority of the bank-of-filters method needs to be analyzed. One factor may be the correlation of the errors among the coefficients for the all-pole mode in the presence of noise. The errors in the elements of a bank of filters are uncorrelated if the noise is white, but this is not true of the errors in the coefficients of a mathematical model. Another factor may be the sensitivity of specific coefficients of the all-pole model.

So, it has been demonstrated for one set of objects that at least two methods for sonar object identification based on the transfer function of the objects will yield a high percentage of correct identifications with a low rate of rejection, provided the signal-to-noise ratio is adequate. The superiority of the uniform bank of filters lends some support to the theory of spectrum analysis of interference patterns although the integration of times are different. The transfer function of an object is represented by the amplitude of the interference pattern across the spectrum. Thus, if neurons can measure peak amplitudes as well as peak spacings (section III), the analysis for object identification is available as a bonus to this processor. The analogy for TSP would be a coloration to the pitch much like timbre in music.

SUMMARY AND CONCLUSIONS

The thrust of this paper has been to present a mechanism of echolocation processing based on spectrum analysis of a signal plus its echoes as well as to present possible methods of object

identification. Justification for this proposed mechanism is based
on the mounting evidence that correlation-echolocation processing
is involved in echolocation and on the need to emphasize an alternative to time-domain correlation techniques. The main strength of
the proposal lies in its universal application to echolocating
animals. This common link applied to the great variety of wideband signals essentially satisfies the goal of an underlying generalization for echolocation that has long been sought. The actual
mechanism of spectrum analysis of interference patterns remains to
be demonstrated. Insight into this processing could aid in the
explanation of known behaviors relating to depth-of-field adjust-
ments and interference rejection, among others.

The object identification techniques presented will yield a
high percentage of correct identifications with a low rate of rejection provided the signal-to-noise ratio is adequate. A method
that is aspect-independent should be developed, although preliminary
investigations suggest that the principle components transformation
extracts invarient features of an object from data obtained at many
aspects (Chestnut and Landsman, 1977). Extensions of these methods
remain to be investigated.

REFERENCES

Altes, R. A., and Reese, W. D., 1975, Doppler-tolerant classifica-
        tion of distributed targets-- a bionic sonar, IEEE Trans.
        on Aerospace and Electronic Systems, AES-11:708.
Altes, R. A., and Titlebaum, E. L., 1970, Bat signals as optimally
        Doppler tolerant waveforms, J. Acoust. Soc. Amer., 48:1014.
Bilsen, F. A., and Ritsma, R. J., 1969, Repetition pitch and its
        implication for hearing theory, Acustica, 22:63.
Chestnut, P., and Landsman, H., 1977, Sonar target recognition
        experiment, ESL Inc., ReportN° ESL-ER178 (Address: 495 Java
        Dr., Sunnyvale, Ca.)
Henson, O. W., Jr., 1970, The ear and audition, in: "Biology of
        Bats", W. A. Wimsatt, ed., Academic Press, New York.
Johnson, R. A., 1972, "Energy Spectrum Analysis as a Processing
        Mechanism for Echolocation", Ph. D. Diss., University of
        Rochester, New York.
Johnson, R. A., 1977, Time difference pitch resolution in humans
        and animal echolocation capabilities, J. Acoust. Soc. Amer.,
        62 (Suppl)(Abstract).
Johnson, R. A., and Titlebaum, E. L., 1972, Range-Doppler uncoupling
        in the Doppler tolerant bat signal, in: "Proceedings of the
        1972 IEEE Ultrasonics Symposium".
Johnson, R. A., and Titlebaum, E. L., 1976, Energy spectrum analysis
        a model of echolocation processing, J. Acoust. Soc. Amer.,
        60:484.

Kay, L., 1967, Discussion to: Neural processing involved in sonar, in: "Animal Sonar Systems: Biology and Bionics", R. G. Busnel, ed., Laboratoire de Physiologie Acoustique, INRA-CNRZ, Jouy-en-Josas, France.

Murchison, A. E., 1979, Detection range and range resolution of echolocating bottlenosed porpoise (Tursiops truncatus), this volume.

Nordmark, J., 1960, Perception of distance in animal echolocation, Nature, 183:1009.

Pollak, G., and Henson, O. W., Jr., 1972, Cochlear microphonic audiograms in the "pure tone" bat, Chilonycteris parnellii parnellii, Science, 176:66.

Pollak, G., and Henson, O. W., Jr., 1973, Specialized function aspects of the middle ear muscles in the bat, Chilonycteris parnellii, J. Comp. Physiol., 84:167.

Pye, J. D., 1967, Discussion to: Theories of sonar systems in relation to biological organisms, in: "Animal Sonar Systems: Biology and Bionics", R. G., Busnel, ed., Laboratoire de Physiologie Acoustique, INRA-CNRZ, Jouy-en-Josas, France.

Ritsma, R. J., 1967, Frequencies dominant in the perception of the pitch of complex sounds, J. Acoust. Soc. Amer., 42:191.

Simmons, J. A., 1970, Distance perception by echolocation: the nature of signal processing in the bat, Bydragen tot de Dierkunde, 40:87.

Simmons, J. A., 1971a, Echolocation in bats: signal processing of echoes for target range, Science, 171:925.

Simmons, J. A., 1971b, The sonar receiver of the bat, Annals of the New York Acad. of Sciences, 188:161.

Simmons, J. A., 1973, The resolution of target range by echolocating bats, J. Acoust. Soc. Amer., 54:157.

Simmons, J. A., Howell, D. J., and Suga, N., 1975, Information content of bat sonar echoes, Am. Scientist, 63:204.

Thompson, R. K., and Herman, L. M., 1975, Underwater frequency discrimination in the bottlenosed dolphin (1-140 kHz) and the human (1-8 kHz), J. Acoust. Soc. Amer., 57:943.

Wightman, F. L., 1973, The pattern-transformation model of pitch, J. Acoust. Soc. Amer., 54:407.

Wilson, J. P., 1967, Psychoacoustics of obstacle detection using ambient or self-generated noise, in: "Animal Sonar Systems: Biology and Bionics", R. G. Busnel, ed., Laboratoire de Physiologie Acoustique, INRA-CNRZ, Jouy-en-Josas, France.

Yost, W. A., Hill, R., and Perez-Falcon, T., 1978, Pitch and pitch discrimination of broadband signals with rippled power spectra, J. Acoust. Soc. Amer., 63:1166.

THE PROCESSING OF SONAR ECHOES BY BATS

James A. Simmons

Departments of Psychology and Biology

Washington University, St. Louis, Missouri 63130 U.S.A.

There are two different types of echolocation in bats, really the two extremes of a continuum of variations in the use of echo information by different species. One type is broadband echolocation, used for producing very acute, multidimensional acoustic images of targets, and it is used by all bats. The other is narrowband echolocation, used for target detection and for acoustic imaging of specific target features related primarily to movement. The results of behavioral (Griffin, 1958; Novick, 1977; Schnitzler, 1978; Simmons, 1977, in preparation; Simmons, Howell, Suga, 1975), physiological (Grinnell, 1973; Henson, 1970; Suga, 1973, 1978), and anatomical (Bruns, 1976; Henson, 1970) investigations reveal the existence of these two distinct modes of biological sonar. The dual nature of echolocation appears in the design of the orientation sounds, in the mechanisms for processing the signals, in the performance of echolocation and the kinds of information gathered, and, to the extent that it is understood, in the evolution of echolocation and bats (Simmons, 1979). Many species actually use a mixture of both extremes in different situations (Simmons, Fenton, O'Farrell, 1979).

These different types of echolocation reflect different emphases on the processing of signals either in the time domain or in the frequency domain. Within the auditory system the use of broadband sonar implies the use of temporal or periodicity representations of echoes, and the use of narrowband sonar implies some reliance on spatially-tuned or "place" representations. Such a duality is basic to any description of hearing (Nordmark, 1978; Siebert, 1973; Wever, 1949), and it is indeed a fundamental aspect of the nature of information in signals. This review attempts to describe echolocation as a phenomenon in the context of hearing and thus constitutes an auditory theory of echolocation mechanisms.

BROADBAND ECHOLOCATION

Peripheral Transformations of Signals

The broadband components of orientation sounds are short (less than 5 msec), usually downward frequency-modulated (FM), and often contain multiple harmonics (Novick, 1977; Simmons and Stein, in preparation; Simmons, Howell, Suga, 1975). Initially they are encoded by the auditory system as arrays of nerve discharges spread in space across the differently-tuned fibers of the auditory nerve (which originate at different places along the basilar membrane), and in time across the duration of the signal (Pollak, Marsh, Bodenhamer, Souther, 1977; Suga, 1973). The signals are first transformed by the external, middle, and inner ears of bats (Henson, 1970; Suga and Jen, 1975). One part of this peripheral transformation of FM signals is a slight stretching due to the frequency-dependent traveling wave responses along the basilar membrane in the cochlea. In _Myotis lucifugus_ the stretching amounts to about 60 µ sec per octave of frequency sweep (McCue, 1969).

Another part of the transformation is half-wave rectification of the acoustic waveform at the hair-cell's interaction with primary auditory neurons (Brugge, _et al._, 1969; Kiang, 1966; Rose, _et al._, 1967; Siebert, 1973; Whitfield, 1978). The hair-cells in the Organ of Corti, which convert mechanical into electrical and finally neural activity, are so connected to neurons that displacement of the stimulus pressure in one direction causes excitation of neurons and displacement in the other direction causes inhibition. The activity of first-order auditory neurons therefore is a half-wave rectified representation of the stimulus at the hair-cells. The discharges of these neurons are locked to the excitatory phases of stimulus waveforms. A certain amount of internal noise is added in the form of jitter of the timing of nerve discharges relative to stimulus cycles (see, for example, Goldstein and Srulovicz, 1977). At present no procedurally definitive measure of this jitter is available; the smallest estimate derived from single-unit recordings is about 40 µ sec RMS jitter. New research suggests that it may be rather smaller, of the order of 10 µsec (D. O. Kim, personal communication).

Central Neural Processing

The time of occurrence of discharges in first-order auditory neurons represents the time at which the FM transmission or

echo sweeps down into the excitatory areas of the tuning curves
for these neurons (Suga, 1973; Suga and Schlegel, 1973). The
intensity of the sound is represented by the number of discharges
following the initial discharge and probably also by the particular population of first-order neurons activated, if recent physiological data correctly suggests that there is an array of different excitatory thresholds for different neurons at any particular
characteristic tuned frequency (Liberman, 1978).

A variety of different temporal response patterns occur in
second-order and higher-order auditory neurons (Suga, 1973).
Two categories of neural processing have been identified that use
the timing of nerve discharges to represent stimulus characteristics. Binaural projections of timing and intensity information
to various midbrain auditory centers probably result in determination of the horizontal direction of a target. The timing of
nerve impulses relative to the occurrence of the stimulus is preserved in the auditory pathways of the brain up to the level of
the inferior colliculus (Pollak, Marsh, Bodenhamer, Souther, 1977;
Suga, 1970). The available evidence indicates that time intervals between sonar transmissions and echoes are recoded spatially,
by place, into delay-tuned neurons (Feng, Simmons, Kick, 1978;
Suga, 1978; Suga and Schlegel, 1973). The timing of nerve spikes
evoked by sonar sounds and echoes represents the target's range
and probably also target structure from timing relationships
within complex echoes (Simmons, in preparation).

The vertical direction of a target may also be represented
in the timing of nerve discharges. Another part of the peripheral transformation of acoustic waveforms is passage of the signal
through the pinna-tragus (external-ear) structures of the bat,
which probably introduce reflections with multiple path-lengths
into the waveform emerging at the tympanic membrane. The acuity
of detection of echo-timing shifts by the bat Eptesicus fuscus
(a broadband echolocator) is about 0.5 μsec, which would be adequate to determine vertical direction from the directionally-
dependent impulse response of the external ear (Simmons, in preparation).

At high levels of the auditory system (the auditory cortex,
for example) spatial arrays of excitation are processed as images
to display a variety of kinds of target information (Suga, 1978).
Some of the target features represented spatially in the bat's
auditory cortex were originally represented by the timing of
nerve discharges in peripheral parts of the auditory nervous system. Higher-order image information describing such target features as motion is also represented in the cortex. For example,
the target's range, which is represented temporally up to the
inferior colliculus and probably spatially thereafter, appears to

be displayed neurally in ways relevant to perception of approaching targets (Suga, O'Neill, Manabe, 1978; O'Neill and Suga, 1979).

Target Images and the Sonar Receiver

Echolocating bats behave as though broadband sonar sounds
and echoes are represented in the sonar receiver by the half-wave
rectified autocorrelation functions of the signals. Fig. 1
illustrates the results of an experiment on detection of jittered
echoes by *Eptesicus fuscus* which demonstrates both the fine temporal acuity of broadband echolocation and the presence of the
equivalent of phase information in the neural representation of
the signals. Ambiguities in perception of echo jitter correspond
to jitter times which are small integer multiples of the average
period (Simmons and Stein, in preparation) of the signals (Fig. 1).
The distinction between the actual autocorrelation function and
its envelope, if no significant smoothing takes place, is not
easily made if the autocorrelation function is first half-wave
rectified.

The results of behavioral studies of perception of target
range, horizontal direction, and the separation in depth of two
surfaces in complex targets (Peff and Simmons, 1972; Simmons,
1973; Simmons, et al., 1974) can be accounted for exclusively as
a discrimination of echo timing information, either between the
emitted sound and subsequent echoes or between samples of the
same echo at the two ears (Simmons, in preparation). Although
other acoustic cues may be present, for example, interaural intensity differences, the psychophysical data available at present
indicate that they are not used.

Fig. 1 shows results that are crucial to understanding the
mechanisms of broadband echolocation. The constellation of
physiological and behavioral data reviewed above suggests that
the broadband sonar receiver of the bat is a neural system operating on information in the time domain. The auditory system of
the bat extracts and displays periodicities in acoustic stimuli,
which for the bat represent such features of targets as range and
multi-planar structure. The data shown in Fig. 1 indicate that
nerve-discharge timing information is combined across the duration of the peripherally stretched, half-wave rectified stimulus
and across the frequencies (cochlear places) in the FM sweeps to
form an autocorrelation-like distribution of the signal's probable time of occurrence.

The bat's cochlea disperses successively-occurring frequencies
in the downward FM sequence to different tuned locations along
the basilar membrane. The result may be to apply each echo frequency to receptors (hair-cells) and primary neurons that are

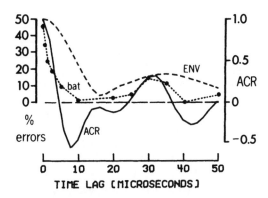

Fig. 1. A graph comparing the performance (% errors; dotted curve with data points) of the bat Eptesicus fuscus at detecting which of two simulated targets is jittering in echo delay (range) with the autocorrelation function of the bat's sonar signals (ACR; solid curve) and with the envelope of the full-wave-rectified ACR (ENV; dashed curve). The bat evidently perceives the half-wave-rectified ACR as the target's image along the delay (range) axis. The region of poor performance for echo jitters around 30 μsec indicates that echo delay changes corresponding to the average period of the signal are less easily detected than jitters of more than or less than one full average cycle (Simmons, in preparation). The curve showing the bat's performance corresponds with what is thought to be physiologically the auditory system's representation of a stimulus.

not very likely to have received much stimulation since the original transmission several milliseconds earlier. The cochlea thus spreads excitation from any one sonar sound or echo across many nerve fibers, setting up a spatial and temporal array of discharges within the volume of the auditory nerve bundle. The auditory nervous system integrates across time and frequency to specify from periodicities in discharges (transmission-echo intervals) an estimated time-of-occurrence for each echo. Fig. 1 shows what psychophysical studies reveal to be the approximate shape of

this estimate as a range or delay image for the stimulus. The
presence in hearing of periodicity-extracting, autocorrelation-
like neural mechanisms has often been hypothesized (Licklider,
1959; Wever, 1949). The intricate psychoacoustical "periodicity"
phenomena can justify theories about both time-domain, periodicity
and frequency-domain, spectral processing operations (DeBoer,
1977; Wightman, 1973). The relatively unambiguous demonstration
of time-domain mechanisms at high stimulus frequencies for broadband echolocation by bats should renew interest in general for
time-domain models of auditory information processing.

It seems less probable that bats use a crosscorrelation
receiver for echolocation than an autocorrelation receiver primarily because the sonar sounds of individual bats change from
one situation to another (Johnson and Titlebaum, 1976; Simmons,
1973). The dimensions along which the signals change to adapt
to different perceptual tasks are generally continuous (Simmons,
Fenton, O'Farrell, 1979). Some species, such as Tadarida brasiliensis (Simmons, et al., 1978), even use radically different
sounds in different situations. Crosscorrelation of transmitted
and received signals requires storage of a replica of the transmission, and bats would require a very large number of replicas,
including not only of the repertoire of emissions but of all
likely Doppler-shifted echoes. An autocorrelation receiver could
detect echoes without formal stored replicas by determining the
presence of many pulse-echo periods of equal length across the
range of frequencies in the FM sweeps of the sonar signal (see
Fig. 1). Of course, the simultaneous presence in the auditory
system of nerve discharges representing the previous sonar emission as well as presently received echoes constitutes a kind of
stored replica. Two forms of the autocorrelation receiver have
been proposed (Johnson and Titlebaum, 1976; Simmons, 1973), a
time-domain autocorrelator and a frequency-domain analyzer of
pulse-echo spectra. Interestingly, the similarity between stimuli
encountered for echolocation of target distance by bats and for
human perception of periodicity pitch suggested these models. The
data reviewed here indicate that time-domain processing is probably used by bats.

The Periodic Structure of Echolocation Signals

It appears likely that one principal function of the bat's
sonar receiver is to extract periodicities from acoustic stimuli
arising from broadband echolocation signals and target echoes to
estimate target features. One characteristic of the orientation
sounds used by most species of bats that are primarily broadband
echolocators is the presence of several harmonic FM sweeps. The
total bandwidth of the signal is large due to the presence of

several harmonic bands, but the signal's periodic structure is
unusual and has interesting perceptual consequences (Simmons and
Stein, in preparation).

Fig. 1 shows the autocorrelation function for an orientation
sound used by Eptesicus fuscus, which we may take to represent
in a general way the orientation sounds of all but the most specialized narrowband bat species (Simmons, 1979). The short,
broadband sonar sounds of Eptesicus contain a strong first-harmonic
sweep from 55 to 25 kHz, a strong second-harmonic sweep from 110
to 50 kHz, and sometimes a weaker third-harmonic in the region of
100 kHz. The harmonic relationship among the frequencies results
in the waveform having an average repeating period equal to the
average period of the fundamental frequency. (The average periodicity is only a useful concept when one is describing closely-
spaced sections of the waveform of a broadband signal such as a
bat sonar sound. The bandwidth is so large that there is sub-
stantial correlation within the waveform only over time intervals
corresponding to a few periods.) The autocorrelation function in
Fig. 1 illustrates the rapid decline in internal correlation by
the sharp decrease in amplitude of the peak at 30 μsec compared
with the central (zero) peak. The average period of the Eptesicus
sound is 30 μsec, not the 14 μsec period corresponding to the
average frequency across all harmonics, which is about 65 kHz.

The decline in autocorrelation for time lags greater than
50 to 100 μsec is almost complete for most broadband bat sonar
sounds (Simmons and Stein, in preparation). This decline is
represented by the width of the envelope of the autocorrelation
function (Fig. 1), or approximately the reciprocal of the signal's
bandwidth, which determines the acuity of target range perception
in many practical situations (Simmons, 1973; Simmons, Howell,
Suga, 1975). The ability of a bat to capture a flying insect in
its wing or tail membrane (Webster and Griffin, 1962) would require
determination of range to within about 2 to 4 cm, which seems
well represented by the envelope of the autocorrelation function.
If the bat must detect a small insect on a surface or resolve the
distinction between different parts of an insect's body for identification of prey, the bat needs all of the time resolving power
potentially available to it. As Fig. 1 shows, this resolving
power corresponds to the half-wave rectified autocorrelation
function of its orientation sound.

To minimize possible interference with perception of small
time intervals by the secondary side peaks in the half-wave-
rectified autocorrelation function (at 30 μsec time lag in Fig. 1),
these side peaks should be kept as far away from the central
peak as possible. The harmonic organization of the bat's sonar
sounds accomplishes such separation by providing lower average

periodicities than would be expected from the average frequency
of the signal as a whole. The signals used by species of bats
that hunt for prey in very cluttered environments are rich in
harmonics, which provides for wide separation of positive-going
peaks in the autocorrelation function. Some species even emit
second and higher harmonics with the lower, first harmonic absent,
often resulting in particularly wide spacing of half-wave peaks
in the autocorrelation function.

NARROWBAND ECHOLOCATION

Bats universally use broadband orientation signals, even if
only as components of sounds dominated by a narrowband component.
When particular species are trained to discriminate target range
or other features requiring fine temporal acuity, the time-resolving
power of their broadband signals is related to performance
(Simmons, 1973; Simmons, Howell, Suga, 1975). When confronted
with perceptual tasks requiring target detection, bats use relatively narrowband signals, judging from observations of the addition of short CF components to signals by Eptesicus (Simmons,
Lavender, Lavender, 1978), or the use of short, almost purely CF
signals during the search stage of insect pursuit by Tadarida
(Simmons, et al., 1978). Many other species emit even longer CF
components and appear to place greater reliance on the kinds of
information conveyed by these very narrowband signals (Schnitzler,
1978).

Peripheral Transformations of Signals

The auditory systems of bats appear to be tuned to the narrowband components of orientation sounds (Grinnell, 1973; Schnitzler,
1978; Simmons, Howell, Suga, 1975; Suga, Simmons, Shimozawa, 1974),
indicating that echoes of narrowband emissions are emphasized.
These CF signals are emitted in situations requiring great sensitivity. Neural responses specifically related to the initial
detection of small targets at ranges likely to be encountered
during hunting have been observed in Pteronotus gymnonotus
(=suapurensis) (Grinnell and Brown, 1978).

The available evidence suggests that short-CF components
(less than 5 msec), which are moderately narrowband signals, are
emitted for detection of targets, and the auditory system is
appropriately tuned to receive a range of biologically relevant
Doppler-shifted echoes of these signals. It is also probable that
detection of short-CF components in echoes involves the use of
mechanisms similar to those involved in processing of broadband
echoes, that is, periodicity detectors. The neural autocorrela-

tion receiver "looks at" the narrowband elements in echoes represented by place in appropriate first-order neurons that are as
sharply tuned as, or perhaps slightly more sharply tuned than,
the first-order neurons tuned to the frequencies in the FM sweeps.
The short-CF echo is also represented by the variety of temporal
response patterns that occur in the auditory pathways to tone
bursts, such as phasic, tonic, "on", and "off" responses (Suga,
1973). Frequency-domain preemphasis occurs in the peripheral
auditory system at those frequencies present in the CF emissions.

Bats that emit very narrowband signals (long-CF components;
more than 5 to 10 msec) extract a variety of kinds of information
from CF echoes that are not available to bats using exclusively
broadband signals or only short-CF sounds. In general terms
long-CF signals convey information about target movements to bats
(Schnitzler, 1978). The peripheral auditory systems of these
bats contain particularly sharply-tuned neurons at frequencies
present in echoes of long-CF transmissions. The frequency of the
CF echo is in fact regulated to the sharply-tuned band by the
bat's Doppler-compensation response. Mechanical tuning at the
periphery in long-CF bats is an order of magnitude sharper than
for other species (Suga, 1978), and resolution of changes in frequency is accordingly very acute (Long, 1977). This improved
peripheral tuning and place representation may be sufficient to
account for the accuracy of tracking at the reference frequency
during Doppler compensation.

Modulations of the amplitude and frequency of long-CF echoes
are represented in a dual manner. They appear as periodicities
in neural discharges (Suga, Neuweiler, Möller, 1976; Suga and Jen,
1977; Suga, Simmons, Jen, 1975). They also appear as represented
by place in the frequency domain since sidebands in the echo
spectrum occur as a result of modulations. Beats produced by
overlap of transmitted and received signals are a kind of modulation in this sense. The average frequency of the CF echo is
represented by the place of activity in different neurons tuned
to different closely-spaced frequencies around the bat's reference
frequency, to which echoes are adjusted through the Doppler compensation response. These same neurons are carrying timing information about echo modulations. The FM components of long-CF/FM
signals, which are the broadband transmissions of these bats, are
represented in the manner described above for broadband signals
in general.

Central Neural Processing

Many features of targets are represented topographically in
the auditory cortex of long-CF/FM bats (Suga, 1978; Suga, O'Neill,

Manabe, 1978). Some of these features, such as average approach
velocity, were initially represented spatially in populations of
first-order neurons as well. Other features, such as target
range, are extracted from temporal information in first-order
neurons, and this presumably is true for periodicities in neural
discharges representing amplitude and frequency modulations of
CF echoes. The acoustically primitive features of targets represented by such straightforward echo parameters as overall intensity, echo delay, and interaural differences in echoes, are represented cortically, as are more complex features compounded from
patterns of information changing over time. Some neural response
patterns appear to be related to specific echo relationships from
naturally-occurring targets. For example, some neurons only
respond to weak echoes at appropriate delays, or to a series of
successive echoes from approaching targets.

Target Images and the Sonar Receiver

The broadband components of both long- and short-CF/FM sounds
probably are processed in the time domain to extract information
for high-resolution acoustic images of targets. The presence of
short-CF components does not so greatly degrade the signal's
bandwidth that these short-CF components must be isolated from
FM components before processing (Simmons, 1973; 1974). There are
no obvious reasons for concluding that the echolocation receiver
is radically different for species using purely broadband, FM
signals than for species using short-CF/FM or FM/short-CF signals
(Simmons, Howell, Suga, 1975). The same periodicity-detecting,
time-domain neural mechanisms presumably serve as the sonar receiver. It may be that the periodicity apparatus in hearing is
a more primitive or evolutionarily more basic auditory system
than the frequency-domain (tuned) or "place" system; comparative
studies of frequency discrimination in a broad range of verte-
brate species support this conclusion (for example, Fay, 1978).

In species of bats where the durations of CF components
exceed 5 to 10 msec and the amount of peripheral tuning and preemphasis for the CF frequencies is great, a qualitatively distinc-
tive frequency-domain mechanism must be present in the receiver.
The Doppler-compensation response of _Rhinolophus_ appears to require
greater frequency resolution and specification of frequency than
can be obtained from beats occurring during pulse-echo overlap.
The physiology underlying fine tuning in the auditory periphery
and the explicit cortical display of average CF echo frequency
reveal the existence of this second, frequency-domain, narrowband
sonar receiver associated with the time-domain receiver in long-
CF bats. Since these bats can detect small modulations in the
amplitudes and frequencies of CF echoes, and since these modula-

tions are coded temporally and by place as spectral sidebands, some combination of time- and frequency-domain processing is likely to occur in the narrowband receiver system. It is possible for a purely frequency-domain (place) mechanism to account for the presently available behavioral data on narrowband echolocation (Schnitzler, 1978), so this question is an important one for future research.

THE ECHOLOCATION RECEIVER

The bat's sonar appears sufficiently different from many conventional sonar-receiver designs to warrant referring to it specifically as the echolocation receiver. Different species of bats seem to have different echolocation receivers, depending upon the extent to which they use narrowband echolocation with CF signals of suitable duration in addition to or in partial replacement of broadband FM echolocation. Several different categories of receivers already appear likely: for example, FM, short-CF/FM or FM/short-CF, and long-CF/FM (Grinnell, 1973; Simmons, Howell, Suga, 1975). Some short- or long-CF sonar systems incorporate the Doppler compensation response (Schnitzler, 1978), while others may not. Significant characteristics of the echolocation receivers of bats are summarized elsewhere in this volume (see reviews by Neuweiler, Pollak, Schnitzler and Henson, and Suga and O'Neill).

Signal Conditioning by the Auditory Periphery

Knowledge about the performance of echolocation and its neural basis suggest that the bat's sonar receiver is organized approximately as illustrated in Fig. 2. This diagram is an auditory model of the echolocation receiver. It shows in a general way the processing of sonar signals and echoes for the target information they contain. The dual, bilateral nature of the auditory system is represented in Fig. 2 (right half of auditory system at the top, left half at the bottom). Furthermore, the apparent unity of the bat's perceptions arrived at through echolocation, a unity revealed in echolocation behavior, is represented by convergence of bilaterally-acquired information upon a single neural display system for multiple dimensions of the acoustic images of targets (right side of diagram). Some means of resolving bilateral rivalry in perception must occur, perhaps through dominance of one of the halves of the auditory system over the other.

The external ears (left side of diagram) act as directional receiving antennae, separated by about 1 to 5 centimeters and

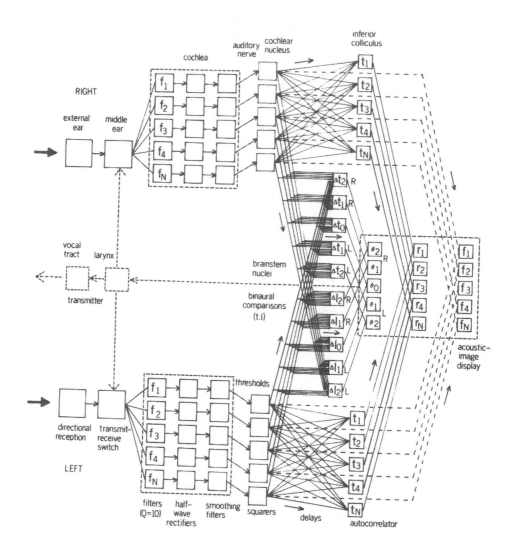

Fig. 2. A diagram of stages known or suspected to occur in the processing of sonar echoes within the auditory system of the bat. This diagram constitutes a model of the echolocation receiver expressed in terms of features of the auditory system (see text). This receiver decodes broadband FM transmission-echo delays and interaural echo time differences to represent target range, multi-planar target structure (as a "fine structure" to range), and the horizontal direction of a target by the place of activity within arrays of neurons. The receiver represents the average approach velocity of a target from average Doppler shifts of narrowband CF echoes by the place of activity in an array of neurons especially sharply tuned to frequencies across the narrow span of biologically relevant CF echo frequencies. The horizon-

pointing forward. The middle ears are the sites of mechanical
switches synchronized to the vocalizations and providing selective attenuation of the transmitted sonar sounds relative to
returning echoes at higher levels in the auditory system (Henson,
1970; Suga and Jen, 1975). Within the auditory system selective
neural attenuation of events representing the transmitted sounds
relative to returning echoes also occurs (Suga and Schlegel,
1973).

The cochlea disperses different sound frequencies to different locations along the Organ of Corti, and, hence, to different
fibers of the auditory nerve. Broadband sonar sounds and echoes
are thus represented in primary auditory neurons by activity in
many parallel channels, each tuned to its characteristic frequency and responding only in the vicinity of that frequency
(f_1 to f_N in Fig. 2). The mechanical and electrical signals
passing through the transduction process within the Organ of
Corti undergo half-wave-rectification and some low-pass filtering (smoothing) prior to triggering nerve impulses in primary
auditory neurons. These events appear to occur within each
tuned channel and are part of the response characteristics of
individual fibers of the auditory nerve. The amplitude response
of primary and many higher-order auditory neurons is similar to
a squaring operation; the neuron's response exhibits a rather
narrow dynamic range and is relatively unchanged (saturated)
over a wide range of higher stimulus amplitudes (Kiang, 1966;
Pollak, Marsh, Bodenhamer, Souther, 1977; Whitfield, 1978).

The frequencies in the FM orientation sounds of bats sweep
downward, in Eptesicus at rates of the order of an octave per
millisecond. In bats the tuning curves of primary auditory

---

tal direction of a target is decoded from interaural amplitude
differences in CF echoes. The temporal structure of CF echoes
(amplitude- and frequency-modulations induced by fluttering targets) is decoded from time intervals between nerve discharges
locked to the phase of the flutter modulations and represented
by the place of activity within other neural arrays. The same
neural mechanisms that decode transmission-echo delays to represent target range in channels tuned to FM echo frequencies decode CF echo flutter modulations in the more sharply-tuned channels tuned to CF echo frequencies. The receiver operates most
generally upon information concerning the timing of echoes and
transforms it into information concerning the location of targets
in space, displaying targets on multi-dimensional maps in the
brain. Other parts of the receiver convert echo amplitude or
frequency to spatial information for these same map displays.

neurons with characteristic frequencies anywhere along the broad
FM bandwidth have sharper high-frequency slopes (260-290 dB/octave)
than low-frequency slopes (110 dB/octave) (Suga and Jen, 1975).
Consequently, the sonar signals sweep down into the response
areas of individual neurons very abruptly. From the moment of
first entry into the neuron's response area, such signals reach
the neuron's characteristic frequency within the time span of
only a few cycles or average periods of the stimulus waveform,
and the signal effectively sweeps down and out of the response
area of the neuron within a short time thereafter. In most instances the discharge of the neuron marks the time-of-occurrence
of the first stimulus cycle to contain appreciable energy at frequencies at or just above the neuron's characteristic frequency.
To what extent does the smoothing filter just prior to the auditory nerve-fiber's threshold-trigger mechanism introduce jitter
or smear the registration of the first effective cycle of the
stimulus by the neuron's discharge? The psychophysical data
(Fig. 2) suggest that such jitter is probably smaller than the
average period (30 μsec) of the bat's FM signals. Measurements
of the true physiological jitter of discharges in peripheral
auditory afferent neurons in bats for FM stimuli should be made
as soon as possible.

Extraction of Target Information from Echoes

The information represented in peripheral auditory neurons
by the time-of-occurrence of successive frequencies in FM sonar
sounds and echoes probably supports a surprising proportion of
the bat's perceptions of targets. "Periodicities," or time-
intervals between separate occurrences of any one frequency, such
as in a sonar signal and its subsequent echo, may be extracted
from the neural representations of these sounds by higher brain
centers (Fig. 2). A neural network that displays transmission-
echo intervals spatially is shown as terminating in the bat's
inferior colliculus (t_1 to t_N). These time intervals and their
spatial display illustrate the fundamental role of the auditory
system in converting acoustic information into information describing targets. A target's range is represented acoustically
in the time delay between the sonar transmission and the echo,
and neurophysiological research clearly demonstrates that a
spatial map of range emerges from the first few levels of the
auditory nervous system. The complex of brainstem and midbrain
auditory nuclei must contain, among other things, the neural
time-to-place transformation processor, a kind of autocorrelator
for periodicities in discharge patterns in the auditory periphery
(Licklider, 1959). Echo delays represent target ranges (r_1 to r_N),
and the spatially presented range-image developed by the auditory system probably undergoes higher-order neural processing in

a manner similar to that observed in the visual cortex.

Interaural comparisons of time (Δt) and intensity (ΔI) are likely to occur at multiple levels in the bat's brain, in the brainstem, midbrain, and auditory cortex. Binaural differences contribute to locating targets in the horizontal plane, and higher neural centers within the auditory system would contain spatial displays of horizontal angular directions (θ) for targets. The neural display of target location in the bat's auditory cortex is discussed elsewhere in this volume (see review by Suga and O'Neill), and neurophysiological data concerning the extraction of information on interaural differences in echoes by the auditory system are also described (see review by Neuweiler).

The relative contributions of interaural time and intensity comparisons to localization of sonar targets are not yet made clear from the results of experiments. The FM sonar sounds of bats are sufficiently broadband that they do provide a basis for discrimination of very small interaural time differences (overcoming the small interaural distances in bats), and the acuity with which bats using broadband sounds actually can detect time changes in echoes is correspondingly fine (Fig. 1). Echolocating bats can in fact discriminate differences in the angles separating targets (see review by Schnitzler and Henson, this volume). One behavioral study of horizontal-angle discrimination by Eptesicus (Peff and Simmons, 1972) yielded results compatible with the use of interaural echo time cues (Simmons and Stein, 1979). Bats may localize targets from broadband echoes by using their external ears as the two antennae of an acoustic interferometer (see Escudie, this volume). A spectral model for the localization of targets by bats using FM sounds requires only interaural intensity discriminations at different frequencies, not interaural time discrimination (discussed comparatively by Grinnell, 1973). The conventional distinction between interaural time (or phase) and intensity cues in sound localization arises from psychoacoustical research on localization of sounds, pure tones in most cases, by humans. A demonstration that individual neurons in the bat's auditory system can respond selectively to tone-bursts with particular interaural intensity differences may only be relevant to understanding target localization with CF echoes. Bats using CF sounds are directionally sensitive to echoes (Grinnell and Schnitzler, 1977), and interaural intensity differences seem likely to provide the basis for this directionality. Interaural time differences have not yet been directly studied as cues for driving the responses of neurons in the bat's auditory system. The distinction made between time and intensity cues in localization does not imply the necessary existence of separate mechanisms for extracting each type of information; binaural crosscorrelation processes may act upon both (see Altes, this volume).

Tonotopic (frequency) mapping is preserved from the cochlea through various levels of the bat's auditory system, and the frequency composition of sounds is displayed in central brain sites (f_1 to f_N in Fig. 2). The neural representation of stimulus frequencies, which originates in the tuning of primary auditory neurons, forms the basis for narrowband echolocation with long-CF signals (see reviews by Neuweiler, by Pollak, and by Suga and O'Neill in this volume). The Organ of Corti contains regions of especially sharp tuning of neurons to the frequencies occurring in CF echoes. The overall or average Doppler shift of echoes is represented by which of these neurons are activated by the echoes, and modulations of these same CF echoes produced by target fluttering motions are represented in part by a temporal or periodic structure to the discharges within these sharply-tuned channels. The same periodicity-extracting mechanisms (t_1 to t_N in Fig. 2) that process broadband echoes for their delay times to display target range (r_1 to r_N) may, in the sharply-tuned channels of "foveation" to CF echo frequencies (see reviews by Neuweiler and by Pollak in this volume), extract modulation rates and display wing-beat frequencies of insects.

The Bat's Description of Sonar Targets

The nature of the acoustic image of a target that bats derive from broadband FM sounds and echoes (Fig. 1) suggests that bats perceive complex targets as a series of "glints" (see reviews by Altes and by Floyd in this volume). A complex target would have a fine structure to its range; different parts of the target would be perceived at slightly different distances. Fig. 1 indicates that Eptesicus can reliably perceive separate glints if their ranges differ by more than 0.1 to 0.2 mm, corresponding to echo delay differences greater than 0.5 sec. Complex targets with angular widths greater than a few degrees would be represented by separate glints localized in slightly different directions. It seems likely that a complex target would appear as distributed across a segment of the bat's cortical spatial map. A target with a surface continuously sloping away in range would reflect a stretched or smeared echo not readily located at a discrete point in range (see Altes, this volume), and the smeared neural registration of the time-of-occurrence of such an echo would be a clue to the presence of the slope. Experiments with complex targets and then with simulated echoes that are acoustically simplified to study the bat's reactions to individual echo features are needed to explore the feature composition of complex perceptions and the perception of patterns of targets by echolocation.

The auditory model illustrated in Fig. 2 supposes that most of the bat's perceptions of target features arise from very acute resolution of the time-of-occurrence of sonar echoes. The bat

samples the echo sound field at two points in space corresponding
to the positions of the external ears, and assembles acoustic
images from the information thus obtained. The principal activity
of the broadband echolocation receiver once targets are detected
is the extraction of periodicities from transmission-echo delays,
with slightly different delays for each glint in a complex target
and slightly different delays and echo intensities at each ear.
It may prove useful to adopt so strongly a time-domain model
because it provides a clear reference point from which to state
alternatives. Purely frequency-domain models have been proposed
(see review by Johnson in this volume), and combinations of time-
domain and frequency-domain processing must also be considered
(Simmons, 1973; Simmons, et al., 1974; see review by Beuter in
this volume). Periodicity models of some auditory phenomena are
strongly favored by comparative auditory evidence (Fay, 1978) and
from many general considerations in auditory theory (Wever, 1949).

Acknowledgments

This research was supported by Grant No. BNS 76-23670 from
the U.S. National Science Foundation. I thank R. A. Altes,
K. Beuter, J. L. Goldstein, R. A. Johnson, and W. A. Lavender for
critical suggestions, and I thank C. Hale, S. A. Kick, and B. D.
Lawrence for assistance in research. I also thank R.-G. Busnel
and the organizers of the Animal Sonar Symposium for the opportunity to present this material.

REFERENCES

Brugge, J. F., Anderson, D. J., Hind, J. E., and Rose, J. E., 1969,
    Time structure of discharges in single auditory nerve fibers
    of the squirrel monkey in response to complex periodic sounds,
    J. Neurophysiol., 32:386.

Bruns, V., 1976, Peripheral auditory tuning for fine frequency anal-
    ysis by the CF-FM bat, Rhinolophus ferrumequinum, II. fre-
    quency mapping in the cochlea, J. Comp. Physiol. 106:87.

DeBoer, E., 1977, Pitch theories unified, in: "Physiology and Bio-
    physics of Hearing", E. F. Evans and J. P. Wilson, eds.,
    Academic Press, New York.

Fay, R. R., 1978, Coding of information in single auditory-nerve
    fibers of the goldfish, J. Acoust. Soc. Amer., 63:136.

Feng, A. S., Simmons, J. A., and Kick, S. A., 1978, Echo detection
    and target-ranging neurons in the auditory system of the bat
    Eptesicus fuscus, Science, 202:645.

Goldstein, J. L., and Srulovicz, P., 1977, Auditory-nerve spike
    intervals as an adequate basis for aural frequency measure-
    ment, in: "Physiology and Biophysics of Hearing", E.F. Evans
    and J. P. Wilson, eds., Academic Press, New York.

Griffin, D. R., 1958, "Listening in the Dark", Yale University Press,
    New Haven, 1974, Dover Publications, New York.

Grinnell, A. D., 1973, Neural processing mechanisms in echolocating
    bats, correlated with differences in emitted sounds, J. Acoust.
    Soc. Amer., 54:147.

Grinnell, A. D. and Brown, P., 1978, Long-latency "subthreshold"
    collicular responses to the constant-frequency components
    emitted by a bat, Science 202:996.

Grinnell, A. D. and Schnitzler, H.-U., 1977, Directional sensitivity
    of echolocation in the horseshoe bat, Rhinolophus ferrum-
    equinum. II. Behavioral directionality of hearing, J. Comp.
    Physiol., 116:63.

Henson, O. W., Jr., 1970, The ear and audition, in: "Biology of Bats,
    Vol. II", W. A. Wimsatt, ed., Academic Press, New York.

Johnson, R. A., and Titlebaum, E. L., 1976, Energy spectrum analysis;
    a model of echolocation processing, J. Acoust. Soc. Amer.,
    60:484.

Kiang, N. Y.-S., 1966, "Discharge Patterns of Single Fibers in the
    Cat's Auditory Nerve", M.I.T. Press, Cambridge, Mass.

Liberman, M. C., 1978, Auditory-nerve response from cats raised in
    a low-noise chamber, J. Acoust. Soc. Amer., 63:442.

Licklider, J. C. R., 1959, Three auditory theories, in: "Psychology,
    a Study of a Science", S. Koch, ed., McGraw-Hill, New York.

Long. G. R., 1977, Masked auditory thresholds from the bat, Rhino-
    lophus ferrumequinum, J. Comp. Physiol., 116:247.

McCue, J. J. G., 1969, Signal processing by the bat, Myotis lucifu-
    gus, J. Aud. Res., 9:100.

Nordmark, J. O., 1978, Frequency and periodicity analysis, in: "Hand-
    book of Perception, Vol. IV. Hearing", E. C. Carterette and

M. P. Friedman, Academic Press, New York.

Novick, A., 1977, Acoustic orientation, in: "Biology of Bats, Vol.
    III", W. A. Wimsatt, Academic Press, New York.

O'Neill, W. E., and Suga, N., 1979, Target range-sensitive neurons
    in the auditory cortex of the mustache bat, Science, 203:69.

Peff, T. C., and Simmons, J. A., 1972, Horizontal-angle resolution
    by echolocating bats, J. Acoust. Soc. Amer., 51:2063.

Pollak, G. D., Marsh, D. S., Bodenhamer, R., and Souther, A., 1977,
    Characteristics of phasic on neurons in inferior colliculus
    of unanesthetized bats with observations relating to mechan-
    isms for echo ranging, J. Neurophysiol., 40:926.

Rose, J. E., Brugge, J. F., Anderson, D. J., and Hind, J. E., 1967,
    Phase-locked responses to low-frequency tones in single audi-
    tory nerve fibers of the squirrel monkey, J. Neurophysiol.,
    30:679.

Schnitzler, H.-U., in press, Die Detektion von Bewegungen durch
    Echoortung bei Fledermäusen, Hand. Dtsch. Zool. Ges.

Siebert, W. M. 1973, Hearing and the ear, in: "Engineering Princ-
    iples in Physiology, Vol. I", J. H. U. Brown and D. S. Gann,
    eds., Academic Press, New York.

Simmons, J. A., 1973, The resolution of target range by echolocating
    bats, J. Acoust. Soc. Amer., 54:157.

Simmons, J. A., 1974, Response of the Doppler echolocation system
    in the bat, Rhinolophus ferrumequinum, J. Acoust. Soc. Amer.,
    56:672.

Simmons, J. A., 1977, Localization and identification of acoustic
    signals, with reference to echolocation, in: "Recognition of
    Complex Acoustic Signals", T. H. Bullock, ed., Dahlem Kon-
    ferenzen, Berlin.

Simmons, J. A., 1979, Phylogenetic adaptations and the evolution of
    echolocation in bats (Chiroptera), in: "Proceedings of the
    Fifth International Bat Research Conference", Texas Tech
    University Press, Lubbock, Texas.

Simmons, J. A., in preparation, Perception of echo phase information
    in bat sonar.

Simmons, J. A., Fenton, M. B., and O'Farrell, M. J., 1979, Echo-
    location and pursuit of prey by bats, Science, 203:16.

Simmons, J. A., Howell, D. J., and Suga, N.; 1975, Information con-
    tent of bat sonar echoes, Amer. Sci., 63:204.

Simmons, J. A., Lavender, W. A., and Lavender, B. A., 1978, Adapt-
    ation of echolocation to environmental noise by the bat,
    Eptesicus fuscus, in: "Proceedings of the Fourth International
    Bat Research Conference", Kenya National Academy for Advance-
    ment of Arts and Sciences, Nairobi, Kenya.

Simmons, J. A., Lavender, W. A., Lavender, B. A., Childs, J. E.,
    Hulebak, K., Rigden, M. R., Sherman, J., Woolman, B., and
    O'Farrell, M. J., 1978, Echolocation by free-tailed bats
    (Tadarida), J. Comp. Physiol., 125:291.

Simmons, J. A., Lavender, W. A., Lavender, B. A., Doroshow, C. F.,
    Kiefer, S. W., Livingston, R., Scallet, A. C., and Crowley,

D. E., 1974, Target structure and echo spectral discrimination by echolocating bats, Science, 186:1130.

Simmons, J. A., and Stein, R. A., in preparation, Acoustic imaging
in bat sonar; the echolocation signals.

Suga, N., 1970, Echo-ranging neurons in the inferior colliculus of
bats, Science, 170:449.

Suga, N., 1973, Feature extraction in the auditory system of bats,
in: "Basic Mechanisms in Hearing", A. R. Møller,ed., Academic Press, New York.

Suga, N., 1978, Specialization of the auditory system for reception
and processing of species-specific sounds, Federation Proceedings, 37:2342.

Suga, N., and Jen, P. H.-S., 1975, Peripheral control of acoustic
signals in the auditory system of echolocating bats, J. Exp.
Biol., 62:277.

Suga, N., and Jen, P.H.-S., 1977, Further studies on the peripheral
auditory system of "CF-FM" bats specialized for the fine frequency analysis of Doppler-shifted echoes, J. Exp. Biol., 69:
207.

Suga, N., Neuweiler, G., and Möller, J., 1976, Peripheral auditory
tuning for fine frequency analysis by the CF-FM bat, Rhinolophus ferrumequinum. IV. Properties of peripheral auditory
neurons, J. Comp. Physiol., 106:111.

Suga, N., O'Neill, W. E., and Manabe, T., 1978, Cortical neurons
sensitive to combinations of information-bearing elements of
bio-sonar signals in the mustache bat, Science, 200:778.

Suga, N., and Schlegel, P., 1973, Coding and processing in the nervous system of FM signal-producing bats, J. Acoust. Soc. Amer.,
84:174.

Suga, N., Simmons, J. A., and Jen, P. H.-S., 1975, Peripheral specialization for fine frequency analysis of Doppler-shifted
echoes in the auditory system of the "CF-FM" bat, Pteronotus
parnellii, J. Exp. Biol., 63:161.

Suga, N., Simmons, J. A., and Shimozawa, T., 1974, Neurophysiological
studies on echolocation systems in awake bats producing CF-FM
orientation sounds, J. Exp. Biol., 61:379.

Webster, F. A., and Griffin, D. R., 1962, The role of the flight
membranes in insect capture by bats, Anim. Behav., 10:332.

Wever, E. G., 1949, "Theory of Hearing", Wiley, New York.

Whitfield, I. C., 1978, The neural code, in: "Handbook of Perception,
Vol. IV: Hearing", E. C. Carterette and M. P. Friedman, eds.,
Academic Press, New York.

Wightman, F. L., 1973, The pattern-transformation model of pitch,
J. Acoust. Soc. Amer., 54:407.

SIGNAL PROCESSING AND DESIGN RELATED TO BAT SONAR SYSTEMS

Bernard Escudié

Signal Processing Laboratory
Institut de Chimie et Physique Industrielles de Lyon
25, rue du Plat   69288 Lyon   CEDEX 1   France

BAT SONAR SIGNALS: COHERENT RECEPTION AND RANGE ESTIMATION

Bat sonar signals cannot be correctly described if the information processor (or receiver) which extracts useful information
from the incoming echo added to noise is not taken into account.
Since the Frascati meeting, many investigators have assumed that
the bat has a coherent receiver, or a matched filter, or a correlator (Mermoz, 1967; Cahlander, 1967; Altes, 1971; Escudié and
Héllion, 1973).   Some results have suggested that the bat's receiver
is coherent and very near the ideal receiver (Van Trees, 1968).
Simmons (1971) and Simmons and Vernon (1971) measured range discrimination ability in bats.   Range resolution can be defined using
the performances of bats to discriminate between two targets very
close to each other.   The major results are shown in Fig. 1:

$$\frac{d}{C_o} = \tau_o \simeq \frac{1}{B} \qquad \begin{array}{l} d = \text{target separation} \\ C_o = \text{sound speed, } C_o \simeq 340 \text{ m/s} \end{array}$$

These agree with the theoretical results on coherent reception.
Many theoretical studies have been derived from a receiver model
(e.g., Johnson, 1972).   All of these models are very near the
matched filter or "correlator", even the "spectral analysis" model
(Johnson, 1972; Licklider, 1959).   These results now suggest that
the bat's receiver can be described (or modelled) as an ideal one
which processes the echoes as a matched filter (Bullock, 1976;
Altes, 1971).   Many of the theoretical results on optimal signals
described in this review resemble the actual signals emitted by
bats.   For the present, this seems to constitute proof.

715

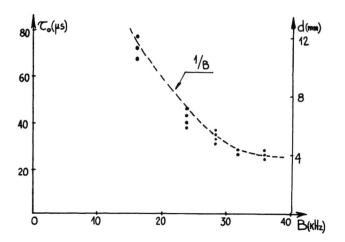

Fig. 1.   Range resolution in bats.

DOPPLER TOLERANT SIGNALS AND ACCELERATION TOLERANT SIGNALS

During cruise and pursuit of prey, bats are faced with the following problem: "How to detect and identify a moving target with constant performances?".

The relative radial speed of the target and the bat, V_R, distorts the signal by the so-called Doppler effect:

$$S(t) \longrightarrow S(\eta t), \quad \eta = \frac{1 - V_{R/C_o}}{1 + V_{R/C_o}}$$
$$t \longrightarrow \eta t$$

The Doppler rate is relatively high (Bard, 1975).

$$C_o \simeq 340 \text{ m/s} \qquad V_R \leq 20 \text{ m/s}$$

$$\frac{V_R}{C_o} \simeq \frac{1}{20} = 5.10^{-2}, \quad 1 \pm 2 \times 5.10^{-2} = 1 \pm 0.1 = \eta$$

Assuming coherent reception with an energy constraint, Altes (1971) derived an optimally Doppler tolerant signal (Fig. 2):

$$S(t) = \frac{A}{\sqrt{t}} J_7(\beta t) \sin\{2\pi\frac{\nu_0}{a}\log(1 + at)\}, \quad T \simeq 3.3 \text{ ms}$$

$$\beta = 2580 \quad J_n(x)n^{th} \text{ order Bessel function.}$$

A similar signal of <u>Myotis lucifugus</u> was recorded by Cahlander (1967) during cruise and pursuit of prey. Other optimally tolerant waveforms have been recorded and analyzed by different authors during recent experiments (Tupinier et al., 1978). Altes (1977) has also shown that the hyperbolic frequency modulated signal processed by matched filter can estimate the relative speed of the target.

The ambiguity function (output of a matched filter) was derived for a high bandwidth duration product BT in hyperbolic frequency modulated signals emitted by bats ($20 \le BT \le 400$)(Bard, 1975). Range estimation seems to be biased by a delay τ_m (Fig. 3).

$$\tau_m(\eta) = \frac{T}{\alpha_m - 1} \cdot \frac{1-\eta}{\eta}, \quad T \text{ duration}$$

α_m is the frequency modulation rate. Tupinier et al. (1978) pointed out that during pursuit:

$$\left| \begin{array}{l} T \text{ decreases} \\ \alpha_m \text{ increases} \end{array} \right. \quad \longrightarrow \quad \tau_m \text{ decreases}$$

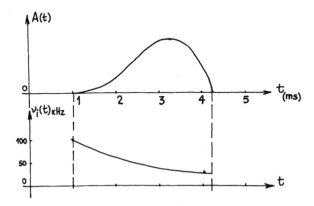

Fig. 2. Amplitude and frequency modulation of a Doppler tolerant
         signal.

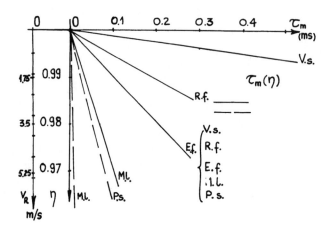

Fig. 3. Bias of range estimation as a function of Doppler effect.

This agrees with previous results reported by Cahlander (1967).

     Taking into account the fact that bats such as Lasiurus borealis
(Cahlander, 1967), Vampyrum spectrum (Bradbury, 1970), or Noctilio
leporinus (Suthers, 1965), emit signals of long duration, Altes
(1971) derived an optimally acceleration tolerant waveform assuming
a coherent receiver. Such a signal is similar to the hyperbolic
frequency modulated cruise signal recorded and analyzed by Cahlander
(1967). These optimally acceleration tolerant signals may be easily
compared with those emitted by different species (Simmons, 1971;
Bradbury, 1970; Suthers, 1965). Assuming a high BT value, the acceleration tolerance of a given signal can be derived using asymptotic
methods (Escudié and Héllion, 1973). Such acceleration tolerance
is very high. Even at acceleration values higher than 10 m/s^2, the
signal to noise ratio loss is less than −3 dB (Escudié and Héllion,
1973).

RANGE AND RANGE ESTIMATION IN BATS: COUPLED AND UNCOUPLED
ESTIMATIONS.

     During target identification and pursuit, the bat estimates
the range and range rate of the flying target. As the speed of the
bat is only two or three times higher than that of insects, the

"pursuit problem" is highly related to the major information: range and relative speed (this is not true for porpoises in hot pursuit of fish!). The incoming echo is mixed with noise; the problem is estimating two different interrelated parameters, range and speed $(r, v_R = \frac{dr}{dt})$. The estimates are correlated and the major problem is reducing the variance of estimation errors (Whalen, 1971):

$$\sigma_{\hat{\tau}}^2 = \frac{1}{E\{\frac{\sigma^2}{\sigma\tau^2} \log p(x|\tau)\}} \cdot \frac{1}{1 - \rho_{\hat{\tau}\hat{\eta}}^2} \qquad \begin{array}{l} \hat{\tau} \text{ range estimate} \\ \hat{\eta} \text{ Doppler rate estimate} \end{array}$$

$$X(t) = S(t) + N(t), \quad (N: \text{ noise})$$

$$\sigma_{\hat{\eta}}^2 \qquad \frac{1}{\{\frac{\sigma^2}{\sigma\eta^2} \log p(x|\eta)\}} \cdot \frac{1}{1 - \rho_{\hat{\tau}\hat{\eta}}^2} \qquad \begin{array}{l} \rho_{\hat{\tau}\hat{\eta}} \text{ correlation rate of} \\ \text{the two estimates} \end{array}$$

The variance $\sigma_{\hat{\tau}}^2$ and $\sigma_{\hat{\eta}}^2$ may be minimized so as to reach the minimum Cramer-Rao limit (Whalen, 1971; Van Trees, 1968), if:

$$\begin{array}{l} 1 - \rho_{\hat{\tau}\hat{\eta}}^2 = 1 \longrightarrow \rho_{\hat{\tau}\hat{\eta}} = 1 \\[2ex] 1 - \rho_{\hat{\tau}\hat{\eta}}^2 \simeq 1 \longrightarrow \rho_{\hat{\tau}\hat{\eta}} \simeq 1 \end{array}$$

Such a solution implies that the designed signal provides uncorrelated estimates for range and range rate (Whalen, 1971; Van Trees, 1968). The signal design problem may be written as follows (Levine, 1970):

$$\lambda_L - <\omega> <t> = 0$$

$$<\omega> = \int_R \pi\nu|\gamma(\nu)|^2 d\nu, \qquad <t> = \int_R t|Z(t)|^2 dt, \quad |Z(t)| = A(t)$$

$$\begin{array}{l} S(t) = A(t)\cos(t) \\ Z(t) = A(t)e^{i\Phi(t)}, \quad BT >> 1 \end{array} \qquad \begin{array}{l} \rightleftharpoons \text{ Fourrier transform;} \\ Z(t): \text{ analytic signal} \\ \text{related to } S(t). \end{array}$$

$$\frac{1}{2\pi} \cdot \frac{d\Phi}{dt} = \nu_i(t): \text{ instantaneous frequency}$$

An easy solution is given by:

$$\begin{array}{l} \nu_i(t) = \nu_i(-t) \\ A(t) = A(-t), \quad -\frac{T}{2} \leq t \leq +\frac{T}{2} \end{array}$$

$$A \text{ and } \nu_i \text{ are even functions!}$$

Even hyperbolic frequency modulated signals have been derived by Altes (1977).

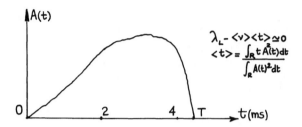

Fig. 4a. Amplitude modulation of quasi-optimal estimation by bats.

Mamode (1979) claims that many bat signals have very low $\rho_{\hat{t}\hat{n}}$ coefficients such that

$$\rho_{\hat{t}\hat{n}} < 0.2 \longrightarrow 1 - \rho_{\hat{t}\hat{n}}^2 \approx 0.96$$

and that some bat signals are very near an optimal solution (Tupinier et al., 1978) (Fig. 4a and 4b). An optimal derivation of such uncorrelated estimation signals may be performed under practical constraints such as those that the bats meet: Doppler tolerance, range resolution, energetic constraints.

ANGULAR ESTIMATION OF TARGETS, ANGULAR IDENTIFICATION AND HARMONIC STRUCTURE

Experiments on <u>Vampyrum spectrum</u> by Bradbury (1970) have shown that bats are able to discriminate between spheres and prolate spheroids (ellipsoids). Some of the signals recorded during identification exhibit many harmonic components (Peff and Simmons, 1972). Simmons noted that under different conditions bats were emitting signals with different harmonic structures (Escudié and Hellion, 1973). Peff and Simmons (1972) recorded signals and performances of <u>Eptesicus fuscus</u> during discrimination of two spheres in different directions.

If we assume that the bat is crosscorrelating the echoes arriving in the two pinnae, as described by many models of hearing (Bard and Simmons, 1976), we can describe the receiver as an "interferometric receiver" whose performance can explain the results obtained by Peff and Simmons, illustrated in Fig. 5 (Escudié and Héllion, 1973). The high frequency harmonic components <u>increase</u> the angular resolution as shown by the correlation function of the signal (Escudié and Héllion, 1973).

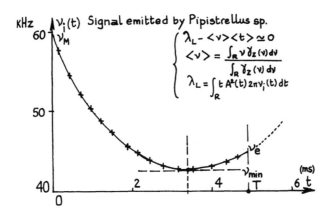

Fig. 4b. Frequency modulation of quasi-optimal estimation by bats.

Bard derived the angular response of such an interferometer without any narrow bandwidth approximation (Fig. 5). The results prove that angular resolution is primarily related to high frequency and low frequency components in the spectrum. Bard and Simmons (1976) have shown that the theoretical results agree with the previous results obtained by Peff during experiments with Eptesicus fuscus (Fig. 6) (Bard et al., 1975, 1977a, 1977b, 1978).

Bard has shown that the harmonic structure of a given signal (e.g., signals emitted by Vampyrum spectrum (Fig. 7)) is a major parameter related to angular response of the interferometric receiver. Furthermore, it may be proved that a "distance frequency equivalence" exists such that bats with small interferometric distances (between the pinnae) are able to emit "wide band" signals to obtain high angular resolution (Bard et al., 1977, 1978).

TARGET IDENTIFICATION AND PARAMETER ESTIMATION

Target Identification

Many experiments using different species (although particularly Eptesicus fuscus) have pointed out the ability of bats to extract target information from the echo. As discussed in the Life Science Research Report 5 (Bullock, 1976):

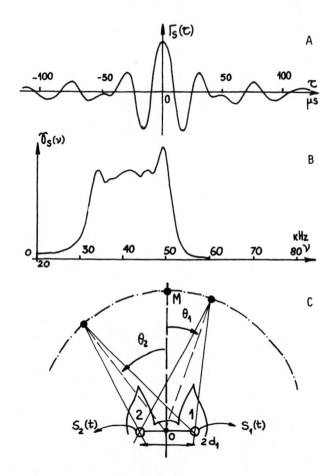

Fig. 5.   Correlation function (A) and power spectrum density (B)
of an <u>Eptesicus</u> <u>fuscus</u> signal.   Interferometric receiver
and targets (C).

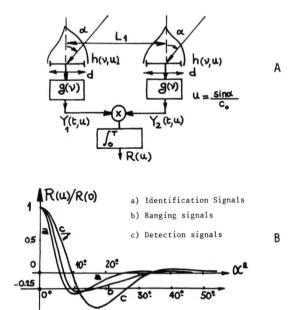

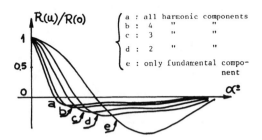

Fig. 6. Bard's Interferometric Receiver Model (A). Interferometric angular response of Eptesicus fuscus signals (B).

Fig. 7. Interferometric angular response of a Vampyrum spectrum signal.

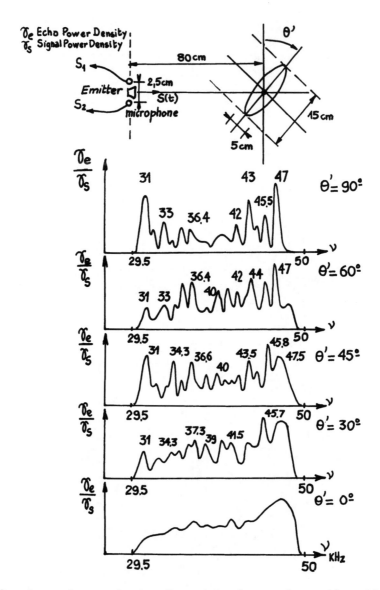

Fig. 8.   Acoustic scattering function of a prolate spheroid.

- Bats can emit "wideband signals" to estimate target range and its size. May we point out the fact that the low frequency components may be used to estimate the shape or volume of the target, as suggested by the "scattering function" or "spectral cross section" defined in Acoustics (Fig. 8).

- Bats emit "wideband signals" with many harmonic components (two or three or more) to estimate small differences in the shape of targets, such as holes (Bullock, 1976). The information content may be described as the complex gain function related to the target. This information requires wideband signals to be extracted by time or frequency processing (Bullock, 1976).

Altes (1975) suggests that a target may be modelled by a set of parallel filters whose fransfer function is related to target parameters.

Assuming coherent reception, many wideband signals emitted by bats can estimate some of these parameters, as demonstrated by Altes and Reese (Bullock, 1976). Even angular estimation provides data on target parameters, as demonstrated using prolate spheroids (Bard, 1977).

Parameter estimation by "narrowband signals" emitted by different bats

Bats such as Rhinolophids or tropical bats (e.g., Pteronotus parnellii) emit "narrowband" signals, called "time diversity" signals by many authors (Escudié et al., 1976). Many examples of such signals have been studied and the frequency modulation may be divided into two parts (Fig. 9):

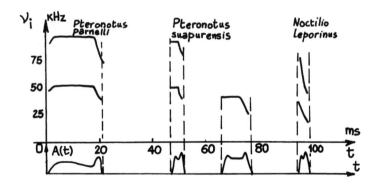

Fig. 9. Amplitude and frequency modulation of "time diversity" signals.

a) Constant frequency or pure tone signals (CF)
b) Frequency modulated signals (FM) which can
    be modulated in a linear or hyperbolic way
    (Suga et al., 1973).

Such signals can be described as "sequential estimation signals".
The first part involves speed or Doppler estimation and the second
range estimation assuming coherent reception (Suga et al., 1973).
Based on Schultheiss and Hill's theoretical approach to sequential
estimation, Escudié and Héllion (1973) pointed out that under opt-
imal conditions (minimum value of $\sigma_{\hat{t}}^2, \sigma_{\hat{R}}^2$) the signals emitted by
Rhinolophus ferrumequinum provide very accurate estimations of tar-
get speed and range.

As the bat's receiver is coherent, such signals can estimate
target speed even in the presence of reverberation. The receiver
processes the signals in the same way as the so-called "Moving
Target Indicator" (MTI) in radar and sonar. Under severe reverberation conditions, the "sub-clutter visibility" (SCV)(which is
the ability to detect the useful echo against the "cluttering reverberation", may reach values close to -15 to -17 dB in the case
of pure tone signals, such as those emitted by Rhinolophus (Escudié
et al., 1976). FM signals whose BT product is close to 20-30 may
be processed in the same coherent way. The signal to reverberation
ratio at the matched filter output is:

$$\left| \frac{S}{R} \right| = 5.4 \, \frac{\sigma^t}{\sigma^c} , \quad BT \leq 40$$

where σ^t is the target cross section and σ^c the cross section of the
scatterers generating the reverberation (Escudié et al., 1976).

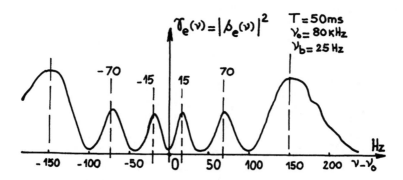

Fig. 10. Power spectrum density of an amplitude and frequency
          modulated echo.

As suggested by Pye (1967) at the Frascati meeting, a theoretical approach of the echo from a flying moth takes into account:

- Periodic amplitude modulation,

- Periodic frequency modulation (highly related to amplitude modulation) of the echo resulting from wing movements (Escudié et al., 1976).

Such a model of the incoming signal demonstrates that the squared modulus of the echo spectrum exhibits "spectral peaks" related to frequency and amplitude modulation caused by target wing movements. These peaks may be analyzed by the bat's receiver (Simmons, 1972), (Fig. 10).

Some aspects of signal processing and signal design related to sonar signals emitted by bats and porpoises are described in a previous review paper (Escudié and Héllion, 1975).

REFERENCES

Altes, R. A., 1971, Methods of wideband signal design for radar and sonar systems, Fed. Clearinghouse N° AD 732-494.

Altes, R. A., 1975, Bionic signal processing for environmental sensing via sonar, electrostatic fields and vision, ESL. TM 637 Rept. Sept.

Altes, R. A., and Reese, W. D., 1975, Doppler-tolerant classification of distributed targets: a bionic sonar, IEEE Trans. on Aerospace and Electronic Systems, AES-11:708.

Altes, R. A., and Skinner, D. P., 1977, Sonar velocity resolution with a linear-period-modulated pulse, J. Acoust. Soc. Amer., 61:1019.

Altes, R. A., and Titlebaum, E. L., 1970, Bat signals as optimally Doppler tolerant waveforms, J. Acoust. Soc. Amer., 48:1014.

Bard, C., Beroujon, M., Escudié, B., Héllion, A., and Simmons, J. A., 1975, Performances de certains signaux sonar animaux à modulation homographique en présence d'harmoniques, in: "Compte-rendu du Cinquième Colloque GRETSI de Nice", B. Derouet, ed., Cagnes-sur-Mer.

Bard, C., Chiollaz, M., Simmons, J. A., and Tupinier, Y., 1977, Etude des performances angulaires de divers systèmes sonar biologiques aériens à l'étude d'un modèle de recepteur interférométrique: imagerie et identification des cibles, in: "Compte-rendu du Sixième Colloque GRETSI de Nice", B. Derouet, ed., Cagnes-sur-Mer.

Bard, C., and Simmons, J. A., 1976 (unpublished), mentioned in: "Recognition of Complex Acoustic Signals", T. H. Bullock, ed., Dahlem Konferenzen, Berlin.

Bard, C., et al., 1978, Acustica, 40, N°3:139.

Bradbury, J. W., 1970, Target discrimination by the echolocating bat, Vampyrum spectrum, J. Exp. Zool., 173:23.

Bullock, T. H., 1976, Life sciences research report, in: "Recognition of Complex Acoustic Signals", T. H. Bullock, ed., Dahlem Konferenzen, Berlin.

Cahlander, D., 1967, Discussion to: Theories of sonar systems and their application to biological organisms, in: "Animal Sonar Systems, Biology and Bionics, Vol. II", R. G. Busnel, ed., Laboratoire de Physiologie Acoustique, INRA-CNRZ, Jouy-en-Josas, France.

Escudié, B., and Héllion, A., 1973, Comparaison entre certain signaux optimaux à grand BT et ceux utilisés par les chauve-souris, in: "Compte-rendu du Quatrième Colloque GRETSI de Nice", B. Derouet, ed., Cagnes-sur-Mer, France.

Escudié, B., and Héllion, A., 1975, Rev. CETHEDEC N° 44, Vol.12:77.

Escudié, B., et al., Rev. Acoustique, N° 38:216. (1976).

Johnson, R. A., 1972, "Energy Spectrum Analysis as a Processing Mechanism for Echolocation", Ph. D. Diss., Univ. Rochester.

Levine, B., 1970,"Radiotechnique Statistique, Vol. II", Mir, ed., Moscow.

Licklider, J. C. R., 1959, "Three auditory theories, in: "Psychology: A Study of a Science", S. Koch, ed., McGraw-Hill, New York.

Mamode, M., in press, in: "Compte-rendu du Septième Colloque GRETSI de Nice", B. Derouet, ed., Cagnes-sur-Mer, France.

Mermoz, H., 1967, Discussion to: Grinnell, A. D., Mechanisms of overcoming interference in echolocating animals, in: "Animal Sonar Systems, Biology and Bionics, Vol. I", R. G. Busnel, ed., Laboratoire de Physiologie Acoustique, INRA-CNRZ, Jouy-en-Josas, France.

Peff, T. C., and Simmons, J. A., 1972, Horizontal-angle resolution by echolocating bats, J. Acoust. Soc. Amer., 51:2063.

Pye, J. D., 1967, Discussion to: Theories of sonar systems in relation to biological organisms, in: "Animal Sonar Systems, Biology and Bionics, Vol. II", R. G. Busnel, ed., Laboratoire de Physiologie Acoustique, INRA-CNRZ, Jouy-en-Josas, France.

Simmons, J. A., 1972, Third International Bat Conference, Plitvice, Yugoslavia.

Simmons, J. A., 1971, Echolocation in bats: signal processing of echoes for target range, Science, 171:925.

Simmons, J. A., and Vernon, J. A., 1971, Echolocation discrimination of targets by the bat, Eptesicus fuscus, J. Exp. Zool., 176: 315.

Suga, N., Schlegel, P., Shimozawa, T., and Simmons, J. A., 1973, Orientation sounds evoked from echolocating bats by electrical stimulation of the brain, J. Acoust. Soc. Amer., 54:793.

Suthers, R. A., 1965, Acoustic orientation by fish-catching bats, J. Exp. Zool., 158:319.

Tupinier, Y., et al., 1978, Analysis of Verpertillionid sonar
        signals during cruise, pursuit, and prey capture, F.I.B.R.C.
        Meeting, New Mexico, U.S.A.
Van Trees, H. L., 1968, "Detection, Estimation, and Modulation
        Theory, Part I", Wiley, New York.
Whalen, A. D., 1971, "Detection of Signals in Noise", Academic Press,
        New York.

MODELS OF SPATIAL INFORMATION PROCESSING IN BIOSONAR SYSTEMS

AND METHODS SUGGESTED TO VALIDATE THEM

Pal Greguss

Applied Biophysics Laboratory
Technical University Budapest
H-1111 Budapest, Hungary

INTRODUCTION

The objective of this symposium is to review the efforts and results in the research on behavior, neurophysiology, ecology, signal processing and electronics engineering with respect to echo-location since the NATO conference on "Animal Sonar Systems - Biology and Bionics" held at Frascati in 1966. I think the Organizing Committee has made a very good decision by choosing 1979 for the year of this symposium. We can now commemorate the 250th anniversary of the man who first described an animal sonar system, that of a bat, the Italian Abbot Lazzaro Spalanzani, who was born on July 12, 1729, at Scandiano in the province of Modena.

Spalanzani discussed his findings first in his correspondence with the Abbot Vassali, as I learned from the first issue of the Archiv für die Physiologie, edited in 1796 by Prof. D. Joh. Christ. Reil, in Halle, Germany (Fig. 1). In Fig. 2, I present a copy of the first two pages of this essay from which we can learn that already at that time, i.e., more than 200 years ago, it was suggested that not only bats but also other animals, and even humans, may have the ability of "seeing" without eyes, if seeing means an information processing, the result of which is an appropriate knowledge of the three-dimensional environment around the species. The answer to this knowledge may then be twofold: orientation and/or target discrimination.

Basically, two types of orientation can be distinguished, depending on whether an organism uses self-generated information carrying waves, or the information carrying wave is independent of the organism. The same distinction holds for pattern recognition

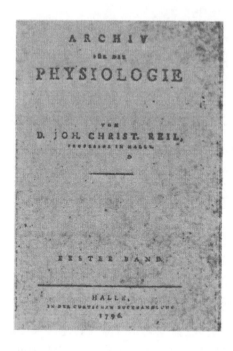

Fig. 1. First page of the Archiv für die Physiologie published
         in 1796, Halle.

too. The difference between the two information processing mechanisms, the active and passive ones, lies in the strategy of the
information handling. In the first case, small time differences
are translated into spatial dimensions, while in the second case,
amplitude (echo intensity) and phase (time) bound information has
to be processed simultaneously.

As a result of animal sonar research all around the world,
synthetic sonars are nowadays probably capable of achieving localization accuracies as good as, and perhaps better than, bats, for instance. However, not only the relative time needed to do so may
significantly exceed that required by these animals, but synthetic
sonars are far beyond the target (pattern) recognition abilities of
animal sonar systems. The reason for this is that we are still
rather far away from understanding the design philosophy which spec-
ifies the neural processing of the received wavefront.

The general sense of conclusion drawn from Simmons' (1979)
review presented at this symposium on the relationship between
behavioral and physiological data on bat sonar and signal processing
theory, Floyd's (1979) comprehensive discussion of the research data

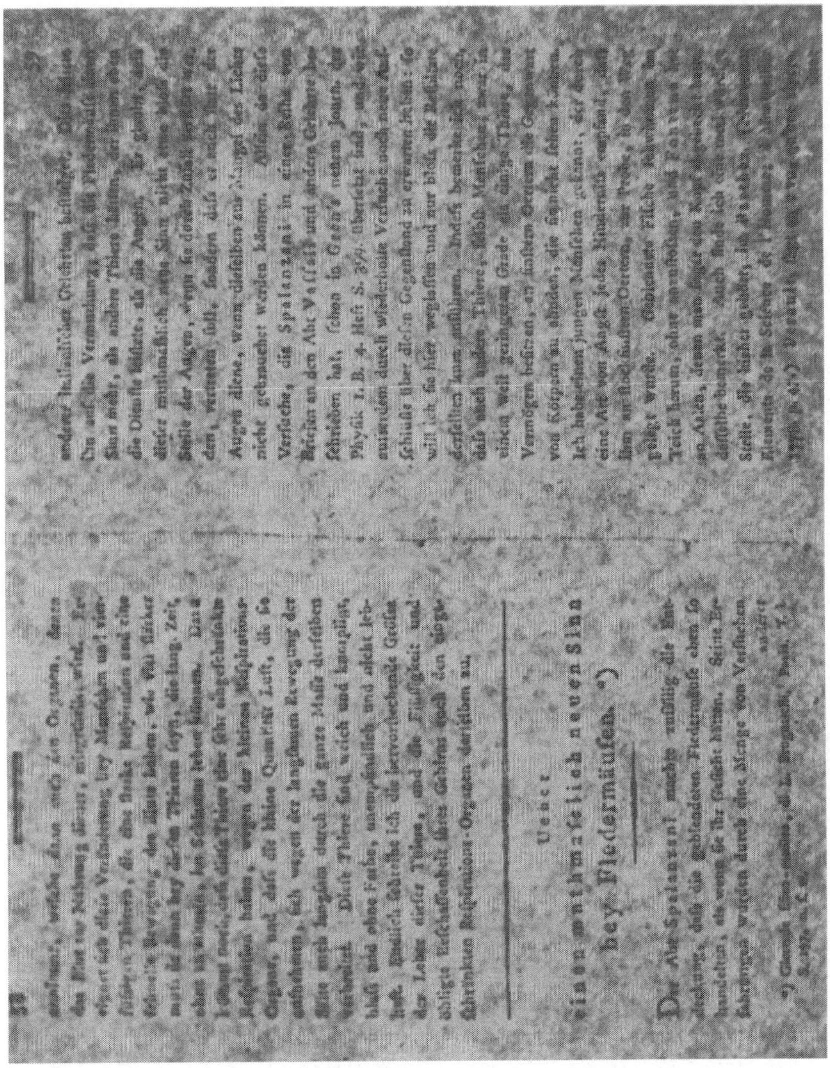
Ueber einen muthmaßlich neuen Sinn bey Fledermäusen. *)

Fig. 2. First pages of the essay "On a Presumable New Sense of Bats" by Brugnarelli.

on cetaceans, and the detailed theoretical work on echolocation by
Altes (1979), is that echolocating animals somehow <u>retain</u> information from the outgoing cry and compare it with information from
returning echoes. Based upon a variety of field and laboratory
results, scientists suggest that there may be a neural "template"
for storing the original emitted wavefront, and some sort of cross
correlation (or autocorrelation) is performed. A number of different hypothetical design philosophies have already been proposed,
each one involving some form of interaction between outgoing wavefront characteristics and the characteristics of the perceived
echoes.

Although these models of animal sonars are capable of describing in a rather conceivable way the functioning of the neural network of echolocating animals where ranging or location is concerned,
they run into difficulty when target and/or shape discrimination
abilities of these animals have to be interpreted. The reason is
simple: the proposed performance strategies are generally not based
upon a <u>simultaneous</u> processing of the amplitude and phase bound
information content of the received wavefront. They presume a
neural activity which only retains the characteristics of the emitted
wavefront, but which is not coherent with the stimulus replica of
the incoming target-reflected wavefront. If, however, we assume
that the stimulus pattern resulting from perceived echo patterns
can interact with a coherent stimulus pattern replica of the emitted
wavefront, then it is evident that the neural network of the animal
is capable of recording the phase bound information of the target
as an intensity (amplitude^2) value simultaneously with its amplitude
bound information, because an interference pattern is formed which
has the properties of a hologram.

The exact meaning of "hologram" is a controversial subject.
Without entering into that controversy, let us give a definition of
holography as an information processing strategy which includes what
is universally called "holography" as one subcase, and many other
methods as additional subcases. According to this definition, holography is an information processing method comprising two stages.
The encoding procedure involves the recording of amplitude and direction of the radiation coming from the target – the resulting interference pattern is the so-called hologram – and the second, decoding,
step involves the direct utilization of amplitude and direction
information it contains. This means that if the object were three-
dimensional, the decoded, reconstructed image will also be three-
dimensional.

Applying this concept to describe information processing strategies in biological systems means that the encoding of the spatial
pattern information is not as generally wave-optical in nature, but
stimulus-like. Both the author of this review (1967) and Dreher
(1967) arrived independently, and practically at the same time, at

the idea of applying the holographic concept to interpret the neurophysiological passage of information in echolocating animals.

The scope of this presentation is twofold. First, it gives a
survey of existing or proposed models of animal sonar systems based
on concepts related to holography. Then, keeping in mind that it is
far more difficult to validate models than to invalidate them, some
methods will be suggested which may prove how far one or another of
the models may match with reality.

MODELS OF AIRBORNE ANIMAL SONARS

Since the letters of Spalanzani, our knowledge about the mechanism underlying the bat's spectacular and highly successful mode of
orientation has considerably increased. We now know that there are
species which use ultrasound of continuous frequency (CF), and others
which emit frequency modulated (FM) signals in which frequency always
sweeps from high to low, while still others use both methods simultaneously: the starting CF signal is terminated with an FM sweep.
It is interesting to note that CF bats are in general less agile
flyers than FM species, which are able to hover in flight, allowing
them to capture crawling or stationary prey.

Considering that the ultrasonic waves emitted by these animals
can be regarded as coherent, it has been assumed (Greguss, 1967)
that the nature of the mechanism that extracts arrival time information from echoes which are reflected and translated into spatial
dimensions is such that the time intervals, i.e., phase bound information, are transformed via a "coherent reference background" in
a reversible way into proportional intensity (amplitude^2) relations
to which the acoustic receptors only are sensitive. It was therefore
predicted that after the wavefront of the target echo enters the
acoustic apparatus, the information about the target (and not only
the range) is transferred into a stimulus pattern, which then proceeds towards the neural organization of the inferior colliculi, the
place where the stimulus pattern replica of the outgoing sound pattern has just arrived, and interferes with it. In other words, both
the amplitude and phase bound information of the target are encoded
in a holographic form in the neural network of the animal. This
prediction was partially based on a remark of Griffin (1950): "the
bat's outgoing pulses can just barely be heard amid the random noise;
the echoes are quite inaudible. Yet the bat is distinguishing and
using these signals some 2000 times fainter than the background
noise!"

This astonishing ability of the bats can be particularly well
interpreted using the above bioholographic model. Fig. 3 shows the
vector diagram of the formation of a stimulus hologram, where R denotes the amplitude of the reference stimulus, I is the amplitude of
the stimulus representing the amplitude of the target echo, and θ

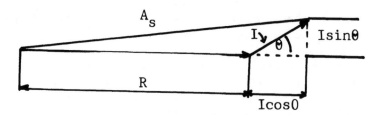

Fig. 3. Vector diagram of the formation of a stimulus hologram.

represents the phase bound information. The resultant stimulus
amplitude, A_s, recorded by the neural network will therefore be

1. $A_s = (R^2 + I^2 + 2RI \cos \theta)^{1/2}$

As seen from Equation 1, R^2 is uniform, I^2 is negligibly small,
and so the information is carried by the last term, which is ampli-
fied by the factor R. This, however, means that when the signal/
noise ratio gets worse, it is more appropriate to increase the value
of R, than to increase the signal strength. In other words, if the
described bioholographic model of echolocating holds, the bat in a
noisy environment will increase the intensity of its inner reference
stimulus pattern, and not the "loudness" of its cry.

Indeed, Airapetianz et al (1967) experimentally demonstrated
that at the time that bats – at least those belonging to the family
Rhinolophidae – emit an ultrasonic impulse, they also send a stimulus
pattern to that part of their brain where the information carried by
the wave reflected from the object is processed, and if the background noise increases, the intensity of this stimulus pattern increases too, without increasing the intensity of the emitted pulse.

To simulate the proposed neural information processing strategy
of Rhinolophidae we desined the model shown in Fig. 4. Although it
worked at higher frequencies than the frequencies ueed by bats, the
medium was water rather than air, which has a higher sound velocity.
The wavelength was therefore about the same order as used by bats.

A piezoelectric transducer P sends a CF wave into a solid block,
the wedge part of which deflects a portion of this wavefront. This
deflected wavefront simulates the emitted CF wave of the bat, which
in turn travels through the environment until it hits the target T,
from which it is reflected. The undeflected part of the wavefront
generated by the piezoelectric transducer represents the stimulus
replica of the outgoing CF wavefront which proceeds towards a definite neural organization of the inferior colliculi denoted by H in
Fig. 4, where it interferes with the stimulus pattern replica of the

Fig. 4. Model of the neural information processing strategy
         of Rhinolophidae.

sound waves reflected from the target T. To investigate how the
intensity distribution of the resulting stimulus pattern - the
biohologram - may look, a sonosensitized plate was placed in plane H
(Greguss, 1966). Fig. 5a represents such an intensity pattern.
That it has the property of a hologram could be proved by reconstruct-
ing from it the image of the insonified target. Fig. 5b shows such
a reconstruction of a structured metal block as a target. The noisy
background is due to the not quite adequate reconstruction technique,
the only one available at that time.

Fig. 5. a) Resulting interference pattern in plane H.
         b) Reconstructed image of a structured metal block.

This information processing strategy may, however, be suitable as an explanation only in the case of CF bats. Bats belonging to the family of Rhinolophidae terminate their CF signal by an FM sweep. The horseshoe bat (<u>Rhinolophus ferrumequinum</u>), for example, emits a CF sound of about 83 kHz for an average duration of 25-50 msec, which is followed by an FM sweep of 1.5 msec duration from 83 kHz down to about 56 kHz. Why does this change occur in the characteristics of the emitted information carrier? Can the holographic concept be extended to that part of the bat's cry?

Our assumption is that in the first stage of their information gathering, the Rhinolophidae are only interested in whether the space volume covered by the emitted ultrasound is empty or whether there are targets in it, and in the latter case, their shape and size, since this knowledge is necessary to be able to avoid them. Those targets which the bat may regard as prey, however, have a size roughly comparable to the wavelength used by the bat, so they cannot get as good an image acoustically as they could optically. How can the bat resolve this obvious inconsistency?

Perhaps by using an information processing strategy similar to that used in optical pattern recognition procedures where the cross correlational properties of the filtering of the Fourier series are used. This, however, presumes that there are neural templates from the prey to be recognized (not identical with the templates of the outgoing cry!) and that these templates (engrams) are stored in a holographic form (Greguss, 1967). If this holds true and the detected sound field contains information in features similar to one of those stored in these templates, a stimulus will be generated, the intensity of which will indicate the degree of correlation. However, this information processing strategy is valid only if the <u>size</u> of the target is the same as of that stored in the engram. Since there is practically no chance that the size of a detected prey will fulfill this requirement, the bat helps itself by terminating its CF signal with an FM sweep.

For a better understanding of this strategy, let us first consider what happens when the bat starts to approach the target, i.e., when it uses CF waves. At this stage, the bat is in the far field of the target, i.e., each point of the target manifests itself as a plane wave propagating in an unique direction. As the bat gets closer to the target, this is no longer valid, and to maintain the former condition the target "has to be put" artificially into infinity. The bat achieves this by changing the "focal length" of its ear which we consider as acting as a varifocal acoustic mirror. If the target gets into the focal plane of the bat's ear, each point on the object is transformed into a plane wave, which is the equivalent of moving the target into the far field.

Dealing now with plane waves, the situation in the receptor system of the bat is somewhat similar to that of an aperture illuminated by a coherent plane wave: a diffraction pattern is formed at a given distance Z from the aperture. Moving a detector along this diffraction pattern with a constant velocity, the intensity of the detector signal will be a function of time. Although no moving sound detector has yet been discovered in the sound receptor system of the bat, it may still experience a similar periodic signal, because sweeping the frequency causes the diffraction pattern to change. If the receptor is fixed at a distance x_0 from the center of the diffraction pattern, and the frequency ν varies with a sweep rate $m = \Delta\nu / \Delta t$, the intensity will vary with time according to:

$$2. \qquad I(t) = \frac{K}{c} mt \left[\frac{\sin (2\pi d/cZ) \, x_0 mt}{(2\pi d/cZ) \, x_0 mt} \right]^2$$

where d is the width of the aperture, c is the sound velocity, and K is a constant which depends on the geometry (Goodman, 1968).

This means, however, that if a coherent receiver is displayed in the hearing system of the bat, it will register an output proportional to the complex field amplitude distribution associated with a portion of the diffracted pattern of the target. It is known from optics literature that a proper coherent processing of data obtained in this way may yield an image of the target, the size of which varies with frequency (Hidaka, 1975; Berbekar, 1975; Farhat, 1975). Since Berbekar and Tökes (1978) have shown for optical pattern recognition that by changing the frequency it is possible to recognize figures of different size, I am almost certain that this holds true for coherent acoustic pattern recognition as well. This then would be the reason why Rhinolophidae terminate their cries with an FM sweep.

I am, however, not unaware that this model is not the only information processing strategy that a bat using an FM signal can pursue, particularly if it is a "pure" FM species. The CF or the CF + FM bats initiate avoidance at around 3 to 10 m distance from the target, while the pure FM bats do so only at around 1 to 3 m distance. This means that pure FM bats have to process information leading to the decision "prey" practically at the same time, which requires most probably another information processing strategy.

MODELS OF UNDERWATER ANIMAL SONARS

Dreher (1967) was the first who considered the possibility that Odontocetes may use some sort of holographic principle in their information processing strategy. He came to this conclusion after analyzing the tegument of the melon of Tursiops. The tegument is anchored to the supporting tissue by structures called rete pegs, and a slice of it shows a remarkably regular array of tiny tubules,

most of them oval in cross section, with an average diameter of
0.01 mm, spaced in longitudinal collumns with an 0.01 mm separation
between columns. Considering the surface as a monostatic receptor
array, he calculated its approximate performance, and found that it
is about 100 lines per millimeter, i.e., in the range of a medium
resolution photographic emulsion. Although this resolution is of
about one order of magnitude less than needed for light holography,
the ultrasonic wavelengths used by dolphins are also several magni-
tudes larger than that of light, so that this resolution capability
matches fairly well the animal's requirement for targets of interest,
usually small to medium fish.

        Apart from mentioning the possibility of a holographic inform-
ation processing strategy, Dreher (1967) did not pursue this idea
systematically. He did, however, refer to an interesting observation
on the behavior of dolphins, quoting Purves (1966), which just may
be a proof of the validity of his idea: there is experimental evidence that a wild Globicephala scammoni may emit a double pulse, as
shown in Fig. 6.

        In general, it is thought that by decoding a single hologram
all information on the object has been reconstructed. This, however,
is not quite true, as shown by Gabor (1965), because during reconstruction one component of the imaging amplitude is dropped, viz.,
the one in quadrature with the background (Equation 1). This loss
of information can nevertheless be avoided if a second hologram is

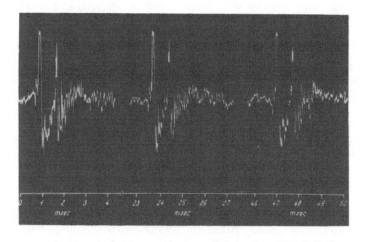

Fig. 6.   Double pulse recorded from wild Globicephala scammoni,
          according to Purves (1966).

recorded from the target with a reference background delayed by
$\lambda/4$, since this will record the component which previously was
dropped. The reconstruction of these two superimposed holograms
will then completely restore the original wavefront, and all information on the target will be available.

The interval between the pulses in these particular signals
seems to correspond to a delay of $\lambda/4$, and so I believe that it is
fair to assume that the reason for such a double pulse is to get
information on the target or prey as complete as possible. My be-
lief is supported by the fact that most CF bats - which most probably use a similar holographic principle - radiate sound from a dipole
source separated so that they can use this principle to reconstruct
all information on targets reaching their processing mechanism.

That blind fish can swim around without touching obstacles
and find their prey was already described by Brugnarelli shortly
after Spalanzani's experiments became known in the late 1790s. Do
they use some sort of sonar system? To answer this question we have
to investigate the role of the lateral line organ which is found
along the sides of the body in all fish and in many amphibians. It
is a special sensory system that perceives vibrations, and which has
many anatomical features in common with mammalian sound receptors.
The mechanical waves generated during swimming are reflected from
different targets and are picked up by the lateral line organ which
produces a stimulus pattern to be processed. The undulating motion
of swimming fish is certainly coordinated by a kind of neural activity, the pattern of which may serve as a reference background in
the information processing strategy of the fish, since it can be
regarded as a stimulus replica of a plane wavefront perpendicular
to the swimming direction. As shown in the sketch of Fig. 7 (for
the sake of simplicity only one reflecting point is indicated), the
reflected waves are spherical, so that the interference of their
stimulus replica with the stimulus template may yield an interference
pattern equal to a one-dimensional zone plate. This means essentially that the information of the given point is coded in a holographic
form, somewhat similar to a "side-looking radar" technique. At
present, this model is only speculative, although Darwin's theory
of evolution allows us to infer the existence of a similar information processing strategy in the neural network of the fish.

Animals using sonar systems live in air or water, i.e., in an
environment in which more phase bound information has to be processed
than on the surface due to the higher degree of freedom of motion.
If an animal lives in an environment where it has to process a lot
of phase bound information, it will most probably use some sort of
holographic principle. If, however, this information is no longer
crucial, the animal may abandon the method. This is just what we
find when we study the sex life of Triturus vulgaris, an amphibian
species.

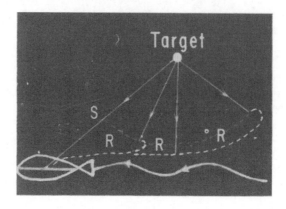

Fig. 7. "Side-looking radar" in fish?

For nine months of the year this amphibian moves on solid surfaces only, but during the three-month breeding season both males
and females return to water. And what happens? Their lateral line
system, which is totally nonfunctional during terrestrial life, re-
develops together with the fin. When returning to an environment
in which living requires more information processing, an organ
evolves which may be used to develop a holographic information processing strategy.

SUGGESTIONS ON HOW TO INVESTIGATE BIOSONAR MODELS

The holographic information processing strategy is only one
of the theoretical models of biological sonar phenomena, and it is
as heuristic as most of them, i.e., it is difficult to test. In
general, the validity of these theories is questioned by a demonstration of the failure of occurrence of requisite event features
at the level of measurement, as was correctly pointed out by Diercks
(1972) in his survey of biological sonar systems. Furthermore, these
models usually fail to predict an unexpected physiological event in
the neural system of the animal. Thus, claims of "solutions" to biological sonar problems are premature and pretentious, and the same
holds true for bioholographic models as well. These claims are
merely statements of the capabilities and limitations of one-dimensional signal form(s). The animal emits and receives patterns of
at least two dimensions, whose structure is barely known at present.
Furthermore, little is known about the acoustic pattern to stimulus
pattern conversion in the receptor apparatus of the animal, although

this knowledge is needed for an understanding of the neural process resulting in a given kind of animal behavior. The purpose of the closing part of my review is to suggest some measuring techniques which may help to validate or invalidate animal sonar models, including of course the reviewed holographic models as well.

The works of Tonndorff and Khanna (1971) introduce the technique of holographic interferometry to study bioacoustical phenomena. From their method of real-time interferometry, researchers have recently switched to the more sophisticated double-exposure interferometry, and it appears now that this technique already offers practical applications in otolaryngology diagnostics. Fig. 8, recorded by von Bally (1977), shows for example the behavior of the human acoustical-to-stimulus converter, the tympanic membrane, at two different frequencies. Why not use a similar method to study the performance of animal sonar systems?

The holographic analysis of the behavior of animal sonar systems would require much care, however, not only because of the size of these systems, but also because of their difficult access. Those who are familiar with optical holographic methods know well that two basic recording configurations exist: the original arrangement of Gabor (1948), where the reference and object wavefronts are in line, and the arrangement introduced by Leith and Upatnieks (1962) where these wavefronts are not collinear, the so-called off-axis method. The merit of the in-line technique is its simple recording

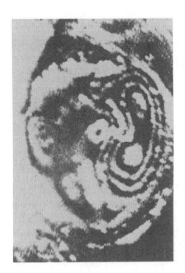 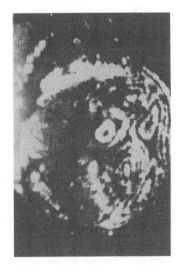

Fig. 8. Vibration mode of the human tympanic membrane at 1.5 kHz (left) and at 3 kHz (right).

configuration. During reconstruction of the hologram, however, the virtual and real images may interfere with each other, and it is rather difficult to separate them. The off-axis method, on the other hand, yields separated images, but it would be rather difficult to use it for testing the behavior of the melon of a dolphin.

To overcome these difficulties, our laboratory is presently developing an easy-to-handle holographic camera based on a forgotten paper of Vanderworker and Snow (1967). The recording begins in the same way as in-line holography, i.e., the laser beam is diverged by a micro-objective, whose optical axis coincides with the axis of the object field. However, between the object and the film (which is placed in the focal plane of the micro-objective) is a beam divider which reflects a portion of the diverging laser beam to the film plane, and this serves as a reference background. Depending on the reflecting properties of the object, beam dividers of different reflecting properties can be used to produce a 2-to-1 reference-to-subject energy ratio, said to be optimum for making holograms. To avoid fogging of a large area of the film by the passage of the laser beam through the photographic film, it is not enough to focus the laser beam at the film plane. An area of about 2 to 3 mm^2 of the emulsion (film) must be removed.

The advantage of this recording technique is that in spite of being in-line during recording, it is off-axis during reconstruction, and rather good reconstructions can be obtained even with white light.

Although the suggested holographic camera may help to furnish ideas on the performance of the acoustical-to-sensory converter of the animal sonar system, it gives us practically no information on the spatial pattern of the acoustic field acting on the acoustical-to-sensory converter (provided that the acting acoustic field is visualized by a method which allows a holographic recording, i.e., the converter does not destroy the coherence of the laser beam). From the growing literature on acoustical-to-optical converters (Greguss, 1975; Wade, 1976; Greguss, in press), I wish to pick out only one, which is based on the change in bi-refrigerant properties induced by an ultrasonic field.

A thin layer of nematic liquid crystals (e.g., MBBA) is sandwiched between two glass plates which are coated with a material that is adherent to their surfaces and yet non-absorbant with respect to the liquid crystals. This allows the crystals to be in a low energy state, aligned on one axis but having no fixed alignments on other axes (Greguss, 1973). An ultrasonic wavefront acting on such a layer can easily produce a change in the predetermined order and state of the layer and consequently alter the optical transmission properties of the system. The liquid crystal molecules are bi-refrigerant, and when they act as wave plates between polarizers, the impinging acoustic wave pattern can be transformed in real time into an equivalent optical pattern. The ultrasonic sensitivity of such a display is at present on the order of several mW/cm^2, so that in theory they could be used to visualize sound fields of echolocating animals.

An advantage of such a display is that a laser can be used without losing its coherence to visualize the acoustic field. Thus, placing such an AOCC display before a holocamera, the spatial pattern of the echo to be produced by the animal sonar system could perhaps be recorded.

REFERENCES

von Bally, G., 1977, Holographic methods in biomedical sciences, Proc. Int. Conf. on Optical Computing in Res. and Dev., 68, Budapest.

Berbekar, G., 1975,"Hologram Aperturak Letapogatasa a Frekvencia Valtoztatasaval", Ph. D. Diss., BME, Budapest.

Berbekar, G., 1978, Hologram aperture synthesis with frequency sweeping, Ultrasonics, 16:251.

Farhat, N. A., 1975, A new imaging principle, in: Proc. IEEE (Lett).

Gabor, D., 1965, Imaging with coherent light, CBS Laboratories Seminar, Stanford.

Gabor, D., 1948, A new microscopic principle, Nature, 161:777.

Goodman, J. W., 1968, "Introduction to Fourier Optics", McGraw-Hill, New York.

Greguss, P., 1966, Pictures by sound, Perspective, 8:287.

Greguss, P., 1967, Bioholography, Lecture at the Society of Electronic Engineers, Budapest.

Greguss, P., 1975, Pictures by sound, Chimia, 29:273.

Greguss, P., in press,"Seeing by Sound", Focal Press, London.

Greguss, P., 1974, USA Pat. 3,831,434. Brevet français, 1973, 2.177.410.

Hidaka, T., 1975, Image reconstruction by using frequency sweep method, J. Appl. Phys., 46:786.

Leith, E. N., Upatnieks, J., 1962, Reconstructed wavefronts and communication theory, J. Opt. Soc. Amer., 52:1123.

Purves, P. E., 1967, Anatomical and experimental observations on the cetacean sonar system, in:"Animal Sonar Systems", R. G. Busnel, ed., Laboratoire de Physiologie Acoustique, INRA-CNRZ, Jouy-en-Josas, France.

Tonndorff, J., and Khanna, S. H., 1971, Validation of holographic observation on the displacement of the tympanic membrane in cats, J. Acoust. Soc. Amer., 49:120.

Vanderworker, R., and Snow, K., 1967, Low spatial frequency holograms of solid objects, Appl. Phys. Lett., 10:35.

Wade, G., 1976, "Acoustic Imaging: Cameras, Microscopes, Phase Arrays and Holographic Systems", Plenum Press, New York.

A NEW CONCEPT OF ECHO EVALUATION IN THE AUDITORY SYSTEM OF BATS

Karl J. Beuter[+]

Fachbereich Biologie, Johann Wolfgang Goethe Universität

Frankfurt am Main, Federal Republic of Germany

1. INTRODUCTION

Bats use their echolocation system very effectively for spatial orientation and prey hunting. Important tasks bats are able to perform are the detection of weak target echoes or echoes superimposed by clutter-interference, the precise localization of targets, and the identification of target structure.

In every echolocation system exist many different requirements which should be fulfilled simultaneously. Basic needs are for instance a high signal energy, a good directionality of transmitter and receiver, and a good time resolution for precise ranging. The analysis of echolocation sounds can give hints how bats have solved this problem in their bionic sonar.

Bat sounds can be recorded and qualitatively analyzed with a rather simple technical apparatus. As a consequence, a variety of different types of sounds has been described. A simple classification scheme according to basic frequency modulation patterns like constant frequency (cf) and frequency modulated (fm) appears more and more unsatisfactory. The goal of theoretical work is to find a generally accepted concept concerning the purpose of the different frequency modulation functions.

A combination of behavioural and systems theoretical methods is the most promising way to find an explanation for the different types of bat sounds. In a given behavioural situation, for example during target discrimination, bats emit very similar sequences of sounds during each approach. Sound sequences found in connection with different target configurations exhibit characteristics, which can be a basis for species specific sound classification.

[+]New address: Battelle-Institut e.V., Am Römerhof 35, Frankfurt am Main, Federal Republic of Germany

747

The purpose of this study is to derive a processing mechanism which does not depend on smaller changes of frequency and amplitude modulation functions. This mechanism should enable the bat to make consistent estimates of target structure, even when the sound pattern is drastically changed as for example during approaching a prey.

Any receiver model should be judged according to its biological relevance. Therefore it has to take into account the basic properties of the neural elements concerned with coding the echo information. The output signals of two basic echolocation receiver types will be dicussed in detail. Ambiguity diagrams and frequency spectra of natural echolocation signals will form the basis for comparing the properties of these receivers.

Since coding of distance and of structural differences in targets has been studied in several behavioural experiments, the resolution of time and time differences in echolocation receivers is of central interest. The echoes of ideal so called "point targets" and simple structured targets will be simulated and analyzed.

The most simple structured target consists of two point targets separated in range by a small distance. To avoid confusion, any target combination will be called a structured target as long as the partial echoes overlap. In this case only one wave train returns to the bat's ear.

Discrimination between structured targets has been studied by Simmons et al. (1974), Vogler and Leimer (1979), Habersetzer and Vogler (1979) in different bat species. The structured targets consisted of massive plates with cylindrical holes of constant depth. The bats were able to discriminate between plates, which had differences in depth as low as 1.5 mm. This corresponds to a difference in sound propagation time of about 8.5 μsec.

Bats can discriminate between structured targets much better than between separated targets. The lower limit for discrimination in range between separated targets is in the order of 15 mm (Simmons, 1973).

2.   THE CONCEPT OF OPTIMUM RECEIVERS

Echolocation systems must derive target information from differences between the emitted signal s(t) and the echo s($\underline{a}$,t). $\underline{a}$ and t represent a set of target parameters and time, respectively. A practically important measure of difference is the root mean square deviation in the form

$$D(\underline{a}) = \int_{-\infty}^{\infty} (s(t) - s(\underline{a},t))^2 dt$$

In the case of energy normalized echoes it follows

$$D(\underline{a}) = \text{const} - 2\chi(\underline{a})$$

where
$$\chi(\underline{a}) = \int_{-\infty}^{\infty} s(t)s(\underline{a},t)dt \qquad (1)$$

In the expression for D, only χ depends on the target parameter $\underline{a}$. Therefore it is sufficient for the receiver to evaluate only χ, which is called the generalized ambiguity function (AF) of the signal s. The AF was introduced in a more special form by Woodward (1953). In a geometrical representation, where signals are described by vectors, the AF can be visualized as a scalar product between the signal- and the echo-vector.

Ambiguity diagrams can serve as a means to describe how a change in a certain parameter influences the deviation of the echo from the expected signal s(t). When the AF changes rapidly with small changes of the parameter, the signal is called "resolvent" for this parameter. If the AF changes only little, it is called "tolerant". Unfortunately it is not possible to make the AF resolvent for all parameters simultaneously. This is due to its so called volume invariance property (Altes, 1973).

Many problems in optimization theory lead to models, the properties of which can be described by the AF. A central problem in echolocation theory are echoes superimposed with interfering noise and ground clutter. Engineers are therefore forced to construct optimal receivers which give the best results according to the following criteria:
-Maximizing of the signal-to-noise ratio at the receiver output.
-Maximizing of the correct signal detection rate at a given "false alarm" rate.
-Optimization of parameter estimation according to a minimum mean square error criterion.
All solutions of this problem (Van Trees, 1968) lead to the same basic signal processing mechanism which is called the "optimum filter", but one has to keep in mind, that it is only optimal for a restricted class of problems. It can be shown that the AF simply describes the output of an optimum filter which is matched to the emitted signal. In technical systems the optimum receiver can be implemented in numerous ways such as correlators, matched filter banks, tapped delay lines or dispersive filters.

Let the target parameter $\underline{a}$ consist only of the Doppler factor $p = (1+v/c)/(1-v/c)$ and the time delay $\mathcal{T} = x/v$, where x is the distance between bat and target, v the relative speed and c = 343 m/sec, i.e. the velocity of sound. This gives the normalized wideband AF in the form

$$\chi(\mathcal{T},p) = p^{1/2} \int_{-\infty}^{\infty} s(t)s(p(t-\mathcal{T}))dt \qquad (2)$$

This type of AF was used in the computer analyses of this study.

Up to this point, signals echoes and filters were only described in

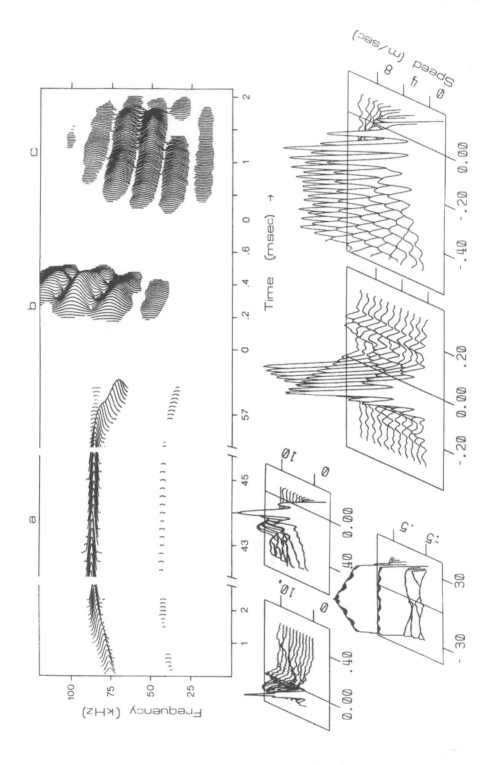

the time domain. The Fourier transform leads to an equivalent description in the frequency domain. Johnson and Titlebaum (1976) pointed out that the energy spectrum contains the same information as the AF, if the Fourier transform is taken from a time interval which contains both the emitted signal and the echo.

Since the output of an optimum receiver gives the best estimate of signal parameters in practically important cases, one can use the AF to derive potential resolving capabilities of bat signals. Energy spectra of signals and their echoes can serve the same purpose.

3.   SOUND RECORDING AND COMPUTATIONAL METHODS

3.1 Animals and Sound Recording

Standard equipment was used to record ultrasonic emissions. Echolocation sounds of four different bat species were recorded in the laboratory when the bats performed defined tasks.

Rhinolophus ferrumequinum and Taphozous melanopogon were flying on a straight path through an empty room of 7x3x2.5 m. Myotis myotis had been trained to catch a mealworm from a stick (Habersetzer, 1978). The sounds were recorded along the whole flight path during approach to the prey. Megaderma lyra had been trained to discriminate between two structured targets (Vogler and Leimer, 1979). The sounds were recorded when the bat was flying 30 cm in front of the targets.

3.2 Computer Analyses and Target Simulations

FORTRAN computer programs written by the author were used on a PDP 11/40 laboratory computer for all signal analyses and for target simulations. After analog-to-digital conversion, bat signals were stored in digital form on magnetic discs. AFs were calculated according to Eq. 2. For graphical reasons and improved visibility, their envelopes were derived by quadrature mixing and plotted in the ambiguity diagrams.

Computer spectrograms which display the frequency modulation functions were constructed from a sequence of short time frequency spectra. The partially overlapping cosine-tapered echolocation sound segments, on which the Fourier analyses were performed, had a duration of 200 μsec, each.

Figure 1 (opposite page): Spectrograms (above) and ambiguity functions (below) of typical echolocation sounds in different bat species. a) Rhinolophus ferrumequinum. b) Megaderma lyra. c) Taphozous melanopogon.

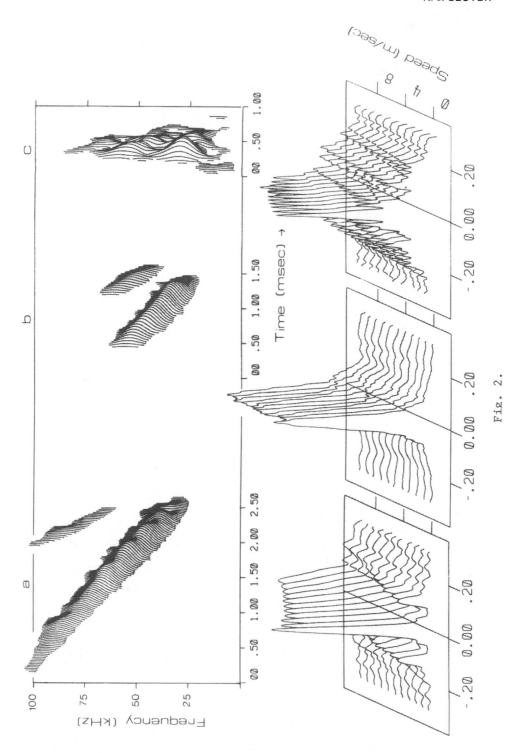

Fig. 2.

Simple structured targets reflecting the sound only in two planes with limited size can be simulated by two point targets shifted in range. The corresponding time shifted signals were added to their original version. The sum was energy-normalized and stored in the same way as normal echolocation sounds.

4.   ANALYSES OF ECHOES

4.1   Ambiguity Functions of Natural Bat Sounds

Typical echolocation sounds from three bat species are shown in Fig.1. The Doppler parameter axis in the AFs is translated into target speed. The shapes of the AFs corresponding to the basic frequency modulation functions cf, cf-fm, short harmonic fm and harmonic cf-fm demonstrate the following general properties.

Cf signals as in the sound of Rhinolophus ferrumequinum allow for a very good speed resolution which can be as fine as 10 cm/sec (Beuter, 1976). Harmonic structures as in the Megaderma lyra sound and those of many other species lead to increased ambiguities, since spurious sidelobes appear. They are caused by the different frequencies simultaneously interfering in the harmonics. The sidelobes are closely spaced and have different hight. Structured targets would produce similar responses. If the bats would use an optimum receiver working only in the time domain, a very precise storage mechanism would be required to compensate for these ambiguities.

In Fig.1a (downward sweep) a cf component at the beginning causes increased Doppler sensitivity, which is expressed in a rapid drift of the maximum of the AF with increasing speed. This leads to an overestimation of distance for approaching targets.

In contrast to the long cf-fm call of Rhinolophus ferrumequinum, Taphozous melanopogon produces fm-sounds with varying sweep rates and many harmonics. The slow sweep rates cause increased Doppler sensitivity and simultaneously the harmonics produce many strong sidelobes. Both effects would call for a very sophisticated evaluation in an optimum processor.

In Fig.2 sounds from three consecutive flight phases of Myotis myotis were analyzed. These types of sound have a very small Doppler sensitivity at least in the biologically relevant speed range below 10 m/sec. The shape of the AF in Fig.2b is very advantageous for precise target ranging and target structure discrimination, since it has only minor sidelobes. A very similar shape of AFs has been found in the hyperbolic fm sweeps of Eptesicus fuscus (Beuter, 1978).

Figure 2 (opposite page): Spectrograms (above) and ambiguity functions of Myotis myotis echolocation sounds during approaching a prey. a) Search phase. b) approach phase. c) Final phase.

When Myotis myotis is approaching a target, the time spread of the
AF increases. The harmonics in the final buzz sounds (Fig.2c) exhibit similar sidelobes as is typical for Megaderma lyra sounds.
From theoretical reasons, the width of the AF along the time axis
is for all signals in the order of magnitude of their inverse bandwidth. Therefore one can expect for all bats with similar sound
bandwidth a width of the AF from 20 μsec to 100 μsec. The high
bandwidth of the long linear fm sweep of Myotis myotis in Fig.1a
is the cause for the very narrowly peaked AF.

Linear or hyperbolic frequency modulations in bat sounds do not
severely influence the shape of the AFs within a biologically significant range of speed. Other unwanted influences such as harmonic structures or irregular sweep rates produce much stronger
effects. This may indicate that there exists no evolutionary
pressure in all bats to produce optimally Doppler tolerant hyperbolic fm sounds as has been speculated by Kroszynski (1968) and
Altes and Titlebaum (1970).

4.2   Echoes From Structured Targets

The output of optimum filters was simulated in the time domain by
caculating envelopes of cross-ambiguity functions between the
actually emitted echolocation sounds and their simulated echoes

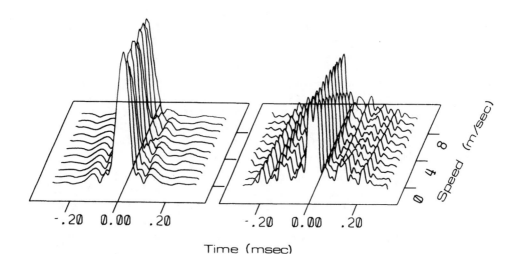

Figure 3: Cross ambiguity function between signals from Figures.
2b and 2c and their simulated echoes from simple structured targets
formed by two point targets separated in depth by 2.8 mm.

from a simple structured (staggered) target with 2.8 mm depth.
Two examples are given in Fig.3 for the sounds from the approach-
and final buzz-phase in Myotis myotis. Separated peaks caused by
the two time-shifted partial echoes can be only observed in the
final buzz sound but the peak separation would lead to an erroneous
depth estimate of about 4 mm. Effects on the amplitude of the optimum filter outputs can also be seen, but they are of minor importance, since in natural situations the bat does not receive energy
normalized echoes and can evaluate the absolute amplitudes therefore only to a limited extent.

Four different representations of the echo of an Eptesicus fuscus
sound reflected from a simple structured target of 14 mm depth are

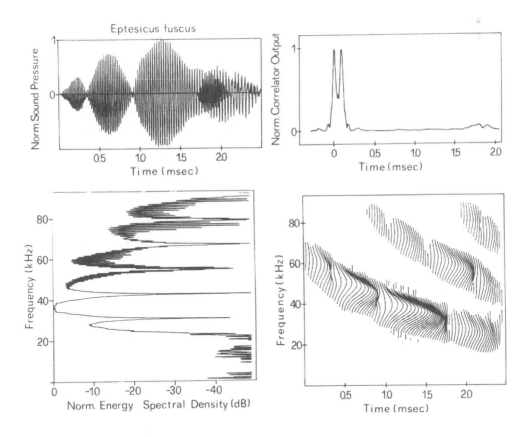

Figure 4: Simulated Echo from a simple structured target formed by
two point targets separated in depth by 14 mm. The original signal
was a natural echolocation sound of Eptesicus fuscus.
Upper left: Waveform. Upper right: Cross-AF between emitted sound
and echo for zero speed. Lower left: Energy spectrum. Lower right:
Spectrogram.

given in Fig. 4. The corresponding echo delay is 82 μsec.

Strong beats appear in the envelope of the echo. They have durations in the order of several hundred μsec. The cross AF for zero
Doppler correctly indicates the two point targets, separated by a
propagation delay of 82 μsec. But the two peaks are only separated
down to about 5o% of their maximal amplitude.

The most strikung effect appears in the energy spectrum, where the
originally smooth curve is periodically depressed. The distance between the periodic minima is 12.2 kHz. The depressions also appear
in the spectrogram at the same frequencies.

The observation can be qualitatively understood on the basis of
interfering waves which are shifted in space by a fixed propagation delay. Destructive interferenc is most effective for those
waves which are shifted by half a wavelength. The effect can be
also derived by simple mathematics. It follows, that the resulting
energy spectrum of the sum of the two echoes contains cosine-
shaped modulations, which are characteristic fot the time delay ΔT
between the echoes. The bandwidth or period of these modulations is

$$B = 1/\Delta T \qquad\qquad (3)$$

This explains the strong modulations in the energy spectrum in Fig.
4. It explains also the beats in the envelope of the echo waveform.
The beats are not regularly spaced as a consequence of the hyperbolic fm and the harmonics. This renders their direct evaluation
difficult.

The narrow oscillations in the energy spectrum are due to the higher
harmonics, which are similar to time-delayed versions of the first
harmonic. These "fast" oscillations in the spectrum are a good
example to visualize the inverse relationship between time delay
and the period of spectral oscillations. For time shifts of 3 msec,
the spectral oscillations have periods of only 33o Hz. To separate
these fine spectral structures, a receiver with its best frequency
at 6o kHz would require a frequency resolution better than o.3%.
At smaller echo delays however, the period of spectral oscillations
increases and is easier to resolve.

5.   A NEW CONCEPT FOR ECHO PROCESSING IN BATS

5.1   Time and Frequency Coding in the Auditory System of Bats

So far only the properties of theoretically derived receivers have
been discussed. We now have to evaluate the coding properties of
the auditory system represented by neural responses.

Suga and Schlegel (1973) have found fm-specialized neurons, which
only respond during a short time interval with very short latencies.

Their potential to dicriminate between a two-tone combination with very small time lags has not yet been measured. Phase-locked responses of single neurons to signal waveforms up to frequencies of 6 kHz have been found. At higher frequencies, the periods of the oscillations could not be distinguished from the time course of the neural responses. The time intervals which correspond to this upper limit of direct temporal coding are longer than 100 μsec.

Recently Vater and Schlegel (1979) have studied the responses of neurons in the inferior colliculus of molossid bats. These bats use strong fm components in their echolocation sounds. A high percentage of phasic on-neurons characterized by very short bursts of activity were found. They respond preferrentially to certain fm stimuli.The short time constants of these phasic neurons make them likely candidates for coding short time intervals down to about 100 μsec.

There is no neurophysiological evidence however, indicating that time intervals far below 100 μsec can be represented by corresponding temporal changes in neural activity. This means that direct coding of changes in the envelopes of echoes or receiver outputs, which are shorter than 100 μsec, is very unlikely.

Processing of information in the frequency domain requires an appropriate representation of signals. The main requirement therefore is a sufficient frequency resolution of the neural elements. Q-10 dB values in fm bats do not differ very significantly from those in other mammals and are typically in the order of 10. A typical neuron with its best frequency at 50 kHz would therefore exhibit a -10 dB bandwidth of about 5 kHz.

5.2 Proposal for a New Processing Mechanism

From the statements above, one can derive a general new concept for structure- and distance-discrimination mechanisms in bats. The model is based on the fact, that the resolution of target structures can be improved, when both temporal and spectral echo representations are analyzed.

The inverse relationship between echo delay and the resulting periods of modulations in theenergy spectrum suggests, that the echo spectrum is used to derive fine structural information. Filtering in the time domain should be used for near optimum detection of weak echoes and for the estimation of larger distances. The proposed mechanism has the following properties.
1. Echo features caused by time shifts of more than a critical time resolution constant are evaluated by direct temporal coding.
2. Echo features which are caused by time shifts smaller than this critical time resolution constant are evaluated in the frequency domain.
3. The critical time resolution constant is in the order of

magnitude of 100 µsec and is similar in all bats.

These three conditions can be simultaneously fulfilled by the same receiver type. Moreover, this receiver can be optimized for typical sweep rates. For instance, the receiver can be constructed from a broadband set of fm specialized bandpass filters with Q-values in the order of 10. Each filter should be followed by an envelope

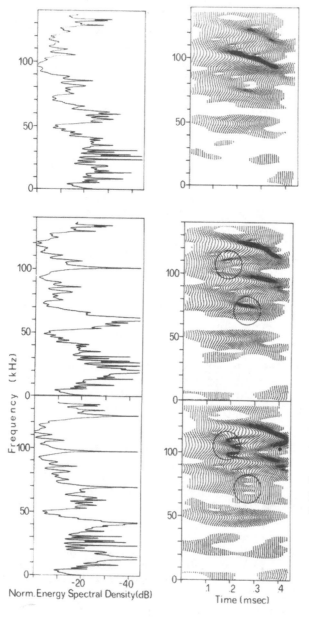

Figure 5: Energy spectra (left) and spectrograms (right) of different computer simulated targets based on a natural echolocation sound of Megaderma lyra.

a) Point target.

b) Simple structured target formed by two point targets separated in depth by 4.1 mm.

c) Target depth of 5.9 mm.

Circles indicate corresponding areas with major differences.

detector which transform the filter output into a sequence of
action potentials.

Temporal features of the echo are coded in this model by the timing
of activity in each channel. Spectral features are represented by
the relative pulse rates in the channels. This coding gives a representation of the echoes in a frequency-time matrix. The distribution
of activity in this neuronal matrix resembles closely to the distribution of the hills in the spectrograms of this study.

Target structure analysis in the frequency-time matrix of this model can be therefore simulated as in Fig. 5. The emitted sound of
Megaderma lyra contains five harmonic downward modulated frequency
bands. The echoes of the structured targets with 5.9 mm and 4.1 mm
depth differ essentially from each other and from the emitted
signal. These differences can serve the bat to discriminate between
the two targets.

Frequency analysis of very short signal segments leads to reduced
frequency resolution in the spectrograms. This inevitable result
of the tradeoff between time and frequency resolution produces the
smooth lines in the spectrograms of Fig. 5. In the proposed model
this effect may be advantageous for processing structural information since random fluctuations in the energy spectra of the
emitted sounds are averaged out. Elimination of variations in the
emitted sounds may lead to improved detectability of target structures.

6. DISCUSSION

Simulated structured targets are well suited to describe the essential properties of echoes from real targets which have been
used in behavioural experiments. This allows for finding out the
shortcomings of optimum processors in their basic form which appear,
when natural echoes are analyzed.

The shortcomings of conventional models can be solved by the proposed mixed model for frequency-time processing. The model can explain the coding of structural information on the basis of presently known neural properties. Extremly fine structural target information can be extracted from spectral echo patterns and range information can be evaluated by direct temporal coding. This method
corresponds to the method for constructing spectrograms. The structural analysis of sonar targets is thus reduced to the recognition
of frequency-time patterns or in the case of tonotopic neural
organization a place-time pattern.

The proposed concept allows for matching the receiver to some
given fm rate and makes optimization of signal-to noise ratio
possible. Especially this receiver mechanism is well suited to
evaluate all types of wideband sonar signals and does not require

receiver adaptations for changing signal structures. The distributed processing makes this model safe against partial failure and gives the bat the opportunity to vary his echolocation call according to external conditions.

7. REFERENCES

Altes, R. A. and Titlebaum, E. L.: Bat signals as optimally Doppler tolerant waveforms. J. Acoust. Soc. Amer., 48, 1014-1020 (1970)

Altes, R. A.: Some invariance properties of the wide-band ambiguity function. J. Acoust. Soc. Amer., 53, 1154-1160 (1973)

Beuter, K. J.: Systemtheoretische Untersuchungen zur Echoortung der Fledermäuse. Doctoral Dissertation, University of Tübingen (1976)

Beuter, K. J.: Optimalempfängertheorie und Informationsverarbeitung im Echoortungssystem der Fledermäuse. In Kybernetik '77, G. Hauske and E. Butenandt eds., Oldenbourg, München-Wien, 1978

Habersetzer, J.: Ortungslaute der Mausohrfledermäuse (Myotis myotis) in verschiedenen Ortungssituationen. Diplomarbeit, Fachbereich Biologie, University of Frankfurt (1978)

Habersetzer, J. and Vogler, B.: Discrimination between structured plates in the bat Myotis myotis (in preparation)

Johnson, R. A. and Titlebaum, E. L.: Energy spectrum analysis: A model of echolocation processing. J. Acoust. Soc. Amer., 60, 484-492 (1976)

Kroszynski, J. J.: Pulse compression by means of linear-period modulation. Proc. IEEE, 57, 1260-1266 (1968)

Simmons, J. A.: The resolution of target range by echolocating bats. J. Acoust. Soc. Amer., 54, 157-173 (1973)

Simmons, J. A., Lavender, W. A., Lavender, B. A., Doroshow, C. A., Kiefer, S. W., Livingston, R., Scallet, A. C. and Crowley, D. E.: Target structure and echo spectral discrimination by echolocating bats. Science, 186, 1130-1132 (1974)

Suga, N. and Schlegel, P.: Coding and Processing in the auditory systems of FM-signal-producing bats. J. Acoust. Soc. Amer., 54, 174-190 (1973)

Van Trees, H.: Detection, estimation and modulation theory, Part I:

Wiley, New York, 1968

Vater, M., Schlegel, P.: Comparative auditory neurophysiology of
    the inferior colliculus of two molossid bats, Molossus ater
    and Molossus molossus. II: Single unit responses to frequency-
    modulated signals and signal and noise combinations. J. Comp.
    Physiol. (in press)

Vogler, B. and Leimer, U.: Discrimination between structured plates
    in the bat Megaderma lyra (in preparation)

Woodward, P. M.: Probability and information theory, with appli-
    cations to radar. Pergamon press, Oxford, 1953.

Chapter VI
Sensing Designs

Chairman : R.S. MacKay

Sensing system design using knowledge of animal acoustic systems.
    R.S. MacKay

Air sonars with acoustical display of spatial information.
    L. Kay

Discrimination of complex information in an artificially generated auditory space using new auditory sensations.
    M.A. Do

SENSING SYSTEM DESIGN USING KNOWLEDGE OF ANIMAL ACOUSTIC SYSTEMS

R. Stuart Mackay

Biology Dept., Boston University

Boston, Mass., 02215 USA

By way of introduction to the topic and the following papers, the term "bionics" might be mentioned. A goal of this activity has been to use the slowly evolved adaptations and mechanisms of living organisms to guide the design of technical developments. There have been some interesting exceptions, but in general this has not often happened. More often the converse has taken place. That is, after a new development has been produced it can be seen that some now understood related systems appear in nature, and in many cases technological progress has removed major mental blocks to understanding animal (and plant) functioning. However, increased learning from nature is becoming possible in some areas.

The natural sonar systems of some bats, birds and mammals, and the related water surface-wave systems of some insects, had come to be understood in many of their aspects largely by comparison with laboriously developed commercial sonar navigation systems (which themselves in some cases are improved by considering more advanced radar systems). Animal performance in these respects has often been excellent compared with what could be achieved by man-made systems as regards speed, reliability, classification capability, jamming freedom, and compact packaging. It should be emphasized that the mere existence of high performance in nature encourages the setting of higher but realistic goals, even if no indication of mechanism is seen...watching birds did not show us how to fly, but suggested we could.

Observation of unexpected performance also forces clarified thinking. For example, Tavolga has found swim bladders of the

catfish *Arius felis* to be direction sensitive to sound at frequencies as low as 100 cycles per second where the wavelength is
about 15 m. One might expect these to sense only a periodic
squeeze from all directions, independent of shape, with emphasis
on no direction. A possible explanation brings out interesting
concepts. Even at large distance, scattering of sound by a very
small low density object is not the same forward and backward,
which creates a "shadow". Thus there is unequal pressure on the
front and back of the object and so it vibrates in the *direction
of sound propagation*. Thus a swim bladder could tug differentially on its supports to give direction sensitivity. The effect
is inherent in equations of Lord Rayleigh from a century ago but
this observation suggests applications to many measuring systems.

Another consequence of related induced motions of gas filled
spaces is that small fish appear significantly larger than actual
to the sonars of whales, making them more noticeable. We use the
same effect to ultrasonically observe and measure the bubbles of
decompression sickness ranging in size down to 1 micrometer with
sound wavelengths of 200 micrometers in animal and human subjects
(Mackay & Rubissow, 1978). (The method mentioned in the next
paper could also be very useful for this application.) In principle dolphins are thereby able to sense things ranging from
dangerously expanding decompression bubbles to a baby requiring
burping.

Of course, the process of making animal observations is
itself instructive and there are reciprocal advantages. Thus
when one understands the observation that a typical hydrophone
records passing airplanes about the same when in or out of the
water of a dolphin tank, one is then better able to consider
animal hearing, and vice versa.

In recent years attempts to understand some aspects of biosonar performances have led to theoretical analyses of such
systems and predictions of their limiting performances. In-
corporation of such thinking into future technology should lead
to improved performance; things have now progressed to the point
that a reverse contribution can be made from understanding
animals to human innovation. Several of the other chapters in
this book contain information against which to measure device
performance, with the attempts to explain and analyze animal
sonars in theoretical terms clearly suggesting approaches to
waveform generation and analysis that could be incorporated into
future generations of man-made systems.

Some general comments can be made. Nature often seems to
employ parallel and peripheral processing of information. This
is widely observed in the visual system but has previously been

little exploited in primary optical systems, though recent
"optical processing" and holographic methods are of this sort.
Various workers contributing to this book observe it in the bat
auditory system. It could be employed with benefit in man-made
acoustic systems now that large scale integrated circuits are
available. (In this connection it is interesting to note that
what has come to be widely known as a "phased array" in the
medical and sonar literature received one of its earliest
mentions in the sound literature at Frascati: Vol 2 pg. 1190.)
In our laboratory we have long routinely sharpened ultrasonic and
other images by mixing with signals their own derivatives, and so
the useful effects of lateral inhibition in all sensory systems
are already understood from technology.

In studying animal systems for mechanisms one often sees
the use of alternative cues. But in attempting to understand
plasticity of function for possible application one must also
consider alternative interpretations of any given set of sensory
system inputs. Thus the ability of a bat to function properly
soon after leaving an altered sound velocity helium-oxygen at-
mosphere is not mysterious if one considers an analogous human
experience. Because the velocity of light in water is 3/4 that
in air, to a diver wearing a usual flat-fronted mask things
appear at 3/4 their actual distance, but the diver is able immediately to reach and grasp objects, probably through sensing
a series of fractional distances and noting closure. Similarly
the reasons for other performances may not follow our precon-
ceived ideas of assumed function. Thus the technology of certain
altimeters emphasized a possible use in distance estimation for
the frequency modulated portion of the cry of a bat. But in experiments with *Tursiops truncatus* I found that animals required
to produce a sustained pure tone of given frequency always preceded it by a sometimes quiet FM sweep, perhaps only as an ad-
justment (Mackay, 1975). In some cases there was also a transient apparently useless following frequency modulation which
training reduced in loudness but did not eliminate. To the
biologist and engineer alike this could emphasize an obvious
caution.

A real contribution of activities such as bionics, or
whatever they are named, is to bring together people from different disciplines who might not otherwise talk to each other.
Significant technical developments as well as biological understanding have come from interdisciplinary thinking. An area
where continuing work has been done that exploits the normal information processing of the animal or human brain, and which
perhaps somewhat mimics the functioning of a two-eared creature,
is in certain ultrasonic blind guidance devices that resolve
multiple object auditory space. There are extensions of applica-

tion ranging from ocean surveillance and fish finding to the
clinical exploration of the movement of the heart in the human
body. The prosthesis application has been one of the most
vigorously pursued areas since our prior Frascati meeting. Two
workers heavily involved in this contribute the next two papers.

References

Mackay, R.S. and Rubissow, G. Decompression Studies Using
    Ultrasonic Imaging of Bubbles. Inst. Electrical and
    Electronics Engineers Trans. on Biomed. Eng. 25, pp. 537-544,
    1978.

Mackay, R.S. Acoustic Control of Environmental Events by
    Tursiops truncatus. In: Conference on Biology and Con-
    servation of Marine Mammals (Abstract), pg. 55. Santa Cruz
    1975.

AIR SONARS WITH ACOUSTICAL DISPLAY OF SPATIAL INFORMATION

Leslie Kay

University of Canterbury

Christchurch, New Zealand

The development of aids to spatial perception for blind persons
has been actively pursued over a period in excess of two decades by
a number of individuals or groups (Benham (6), Kay (23), Benjamin
(7), Russell (42), Collins (15), Armstrong (2), Brindley (11) and
Dobelle (16), to mention only a few who are well known researchers
in the field. The primary goal in the late 50's and the early 60's
was to improve mobility in the blind through the use of a sensory aid.
Now most aids are designed to complement the Long Cane (36) which
emerged during the 60's as a primary mobility aid, following its
introduction in the U.S.A. in 1947. This was partly brought about
through the strong influence of the teachers of orientation and
mobility for the blind, whose teaching skills are based on Long Cane
travel. This demands reliance on the natural sense of hearing as
well as the probing of the cane ahead of the user. Thus, whilst
the design of a sensory aid remains the prerogative of the electronic
engineer it is most important that he pays adequate cognizance to
the experience and views of these teachers before allowing his
fertile imagination free rein. It is against this background that
three distinctly different sensory aid philosophies have developed.

The most elementary philosophy calls for the "clear path
indicator" or obstacle detector. Such an aid to travel by a blind
person does not indicate where the path is. The aids of Benham,
Benjamin, Russell, and Armstrong fit the philosophy. These aids
indicate only the presence of the nearest object in the travel path
and are intended to be easy to use. Clearly they can also be used
to locate objects nearby. All are narrow beam devices and may be
in the form of a torchlight, fitted into a cane, or mounted on the
chest. Russell and Armstrong use ultrasonic waves, whereas
Benham and Benjamin use light waves to gather environmental

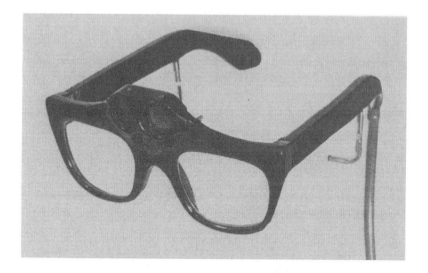

FIG.1.   Sonic Spectacles for the Blind as used in the 1971-72
           Evaluation Programme

information. The display of distance information is in either
auditory or tactile form.

     The second philosophy due to Kay, proposes the use of the
auditory sense to its maximum capability in sensing the complex
environment in which we move. This has led to devices in the form
of a hand-held torch (25), head-mounted in spectacle frames (26) or
a headband as shown in Figs.1· and 2. The display may be either
monaural for a narrow high resolution field of view, or binaural for
a wide field of view. This latter may be designed to match the
auditory system localisation function (46) or be used to give mainly
left, right or ahead indications of the direction of an object as
well as the distance to an object. A special feature of this display
is the means to locate more than one object at a time and the audio
signal carries information about the nature of the object (24).
(This is missing in the "clear path indicators").

     The third philosophy seeks to model vision by displaying frontal
images of the environment using either a matrax of tactile
stimulators on the body (Collins and Bach-y-Rita (4)), or a matrix
of neural stimulators implanted in the brain as proposed by

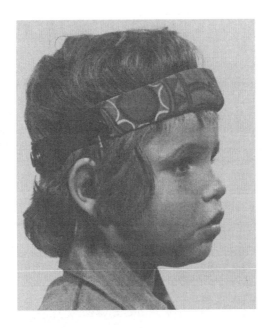

FIG.2.   Spatial Sensor for Blind Children Mounted in Cosmetically
           Acceptable Headband.

Brindley (11) and Dobelle (16). The information is collected by a
miniature camera using solid state microelectronics. At present
the displayed information is elemental (in a visual sense) resembling
at best spots of light (or phosphines), or a 20 x 20 array of
vibrators on the abdomen.

     This paper discusses the theory, design and use of the sonar
system due to Kay which aims to stimulate the auditory sense to its
maximum capability. As such, it may be said to most nearly model
the animal sonar. The "clear path indicators" cannot be considered
in this context except in a simple way, since the environmental
information is so processed as to be reduced to single object
distance only with a stimulus of predetermined form. A comparison
is made with echo-location used by animals and the spatial
perception they might enjoy.

     It has been found through a comparison between continuous

transmission frequency-modulation and tone-burst forms of radiated ultrasonic energy that the former provides a means for matching to the human auditory process (24). The linear frequency sweep can be specially chosen to match optimally the auditory process so as to provide a display of spatial information which is rich in object character and sensitive to motion. User experience has shown that the need for accurate judgment of distance and direction is less important than perception of the flow of spatial sound patterns and locomotor control in negotiating a complex environment can be readily achieved - even by blind children - using only the acoustic output information from an air sonar. It is also possible to observe the motion of objects about their axis of rotation with an appreciation of what is taking place which challenges some high resolution pulse sonars with visual display

There are two possible reasons for this: a) the linear frequency-modulation radiation which covers a frequency band of approximately one octave has an ambiguity function which combines time delay and Doppler shift in a way that a wide band "tone" burst cannot, and b) auditory and visual perception of the display of spatial information are different processes and the motion of sonic reflectors which may be readily perceived auditorily may not necessarily be readily perceived on a visual display.

BASIC CONCEPT OF AUDITORY SPATIAL PERCEPTION

In order to perceive objects in the environment and be aware of their relative motion through the medium of an artificial sensing system it is first necessary to establish spatial coordinates. These are most readily understood as DISTANCE and DIRECTION and the more precise the resolution the greater the environmental detail which can be perceived. This requires a sensing wave energy with as short a wavelength as practicable and a system bandwidth as wide as possible. When vision is impaired as in blindness, the most effective remaining spatial sense is hearing which requires an acoustical form of coupling of the sensor to the ears and a coding of the spatial information which is most readily interpreted. It has been argued that the most effective spatial sense is the tactile sense where spatial information is mapped in two dimensions on the abdomen (15). However, it is later shown here that the auditory sense can handle superior information.

The optimum spatial code appears to have two components. Distance is best represented proportionally by the frequency of the audible signal and directional angle by the proportional logrithmic increase in the interaural amplitude difference (IAD) of sound fed to both ears. Left is indicated by a louder sound in the left ear which causes a lateral shift of the binaural sound image to the

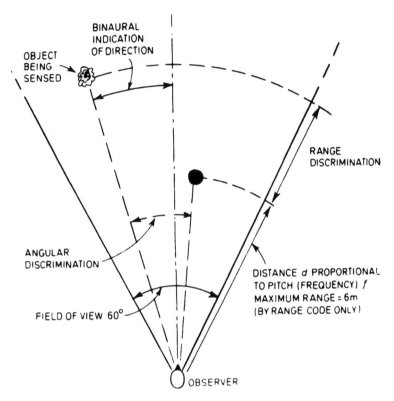

FIG.3.   Illustration of Distance and Direction Code for Spatial
Sensor.

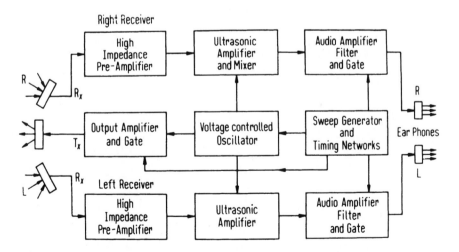

FIG.4.   Schematic of Binaural Spatial Sensor using Ultrasonic
Radiation.

left of the midline. This code may be easily modelled in a computer
for simulation purposes (12) (17), but is not physically achievable
without introducing auditory artifacts which may be disturbing to
the user. Some apparent artifacts are nevertheless complex useful
spatial information which has been found difficult to simulate, as
shown later.

Ideally it would appear that an object acting as a point
reflector of wave energy in the sensing field would cause a
continuous tone to be heard which would be imaged in the head as a
pure tone, the frequency of which was proportional to the distance.
Its position would be shifted laterally in the head by an amount
corresponding to the direction of the object. If the object were
caused to move radially relative to the observer the tone would
decrease as the object came closer and the image would shift
laterally to the left as the object moved to the left.

The closest it is possible in practice to approach such a
spatial code is through the use of a linear ramp modulation on the
frequency of ultrasonic radiation and the reception of echos by
two receiving apertures having divergent polar receiving
sensitivities; each having a Gaussion variation in angular response
relative to the angles of divergence (42).

The code is illustrated in Fig.3 and a schematic of the system
is shown in Fig.4.

PHYSICAL REALISATION OF THE SPATIAL SENSING CODE

Distance Code

The radiated signal of a linear ramp frequency modulation
echo-location system may be represented by the relationship

$$S_T(t) = \exp j\theta_T(t) \qquad (1)$$

where

$$\theta_T(t) = 2\pi \left| f_2 t_s - \frac{m}{2} t_s^2 \right|$$

and

$$t = t_n + t_s,$$ shown in Fig.5.

(t_n is the beginning of the nth cycle of sweep period T_s).

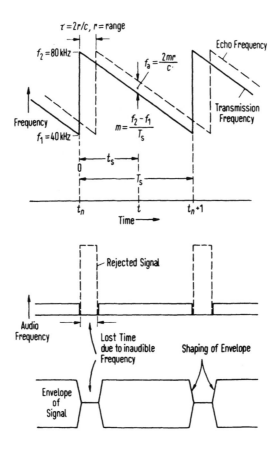

FIG.5.   Frequency Sweep and Audio Signal of Sonar.

Thus, while t varies between t_n and t_{n+1}, t_s varies from 0 to T_s, $T_s \gg \tau_{max} = 2r_{max}/c$

m is the sweep rate in Hz/s , and may be either positive or negative. In the system being described m is negative.

The instantaneous transmitted frequency is

$$f_T(t) = f_2 - mt_s \qquad (2)$$

f_2 is the frequency when $t_s = 0.$

When, due to motion, the received signal at t_s is a Doppler shifted "replica" of the transmitted wave radiated at $t_s - \tau$ the phase angle of the received signal at the time t_s is

$$\theta_R(t) = 2\pi \left\{ f_2 [t_s - \tau(t)] - \frac{m}{2} [t_s - \tau(t)]^2 \right\} \qquad (3)$$

The received frequency is

$$f_R(t) = \frac{1}{2\pi} \frac{d\theta_R(t)}{dt}$$

$$= f - f_2 \dot{\tau}(t) - m [1 - \dot{\tau}(t)][t_s - \tau(t)] \qquad (4)$$

where the dot represents the time derivatives. The distance to the reflecting structure is then indicated by the difference frequency obtained by simple multiplication between $S_T(t)$ and $S_R(t)$ giving

$$f_T(t) - f_R(t) = f_a(t)$$

$$= m \tau(t) [1 - \dot{\tau}(t)] - \dot{\tau}(t) (f_2 - mt_s) \qquad (5)$$

The general expression for the time delay is

$$\frac{\tau(t)}{2} = \frac{1}{c} r \left(t - \frac{\tau(t)}{2} \right) \qquad (6)$$

where $r(t)$ is the instantaneous distance and $\dot{r}(t)$ is positive for increasing $r(t)$.

Since the rate of change of $r(t)$ is much smaller than the velocity of propagation c, the following approximation may be used

$$\tau(t) = \frac{2r(t)}{c} \quad \text{for} \quad \left| \frac{2\dot{r}(t)}{c} \right| \ll 1$$

then

$$f_a(t) \doteq \frac{2m}{c} r(t) - f_T(t) \frac{2\dot{r}(t)}{c} \qquad (7)$$

illustrated in Fig.6.

When the Doppler shift is zero for stationary reflections

$$f_a = \frac{2mr}{c} \qquad \frac{2m}{c} \quad \text{is called the distance code in Hz/m.}$$

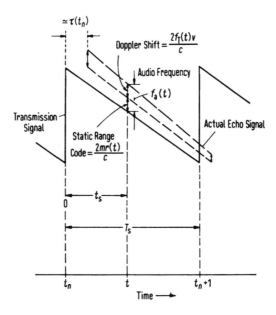

FIG.6.   Doppler Shift Due to Motion.

It should be noted that the term $f_T(t)$ $[2\dot{r}(t)/c]$ is not the Doppler shift induced by the moving structure. It is instead a function of the transmitted frequency at the instant $t = t_n + t_s$ (which is the instant of reception) and the instantaneous distance of the structure at the same time t. The relation is sufficiently general to include non-constant velocity and is in this form to facilitate ease of computation as well as ready appreciation of the range indicator $f_a(t)$.

When r(t) is increasing, i.e. the structure is moving away from the transducers thus making $\dot{r}(t)$ positive, the difference frequency is given by equation (7) as written

The audible signal

$$S_a(t) \; = \; 0 \qquad\qquad\qquad (t_n < t < t_{n+\tau})$$

$$= \; \beta(r,t) \; \cos 2\pi f_a t \; (t_{n+\tau} < t < t_{n+1}) \qquad\qquad (8)$$

$\beta(r,t)$ is both range and frequency sensitive (which is varying with t) carrying information about the reflection characteristics of the object during each frequency sweep.

$S_a(t)$ may thus be expressed as a Fourier Series (24)
for stationary conditions only by

$$\beta(r,t) \sum_{p=1}^{\infty} a_p \cos p\,\omega_s t$$

where $\omega_s = \dfrac{2\pi}{T_s}$

and $a_p = \dfrac{\sin(\omega_a + p\omega_s)k}{(\omega_a + p\omega_s)} + \dfrac{\sin(\omega_a - p\omega_s)k}{(\omega_a - p\omega_s)}$

and is illustrated in fig.7. With motion the line structure is
broken up. This echo signal is heard as an interrupted tone, the
quality of the tone being determined by the ratio T_s/τ which
should be as large as practicable. A high value of ω_s (say 25)
causes a faint motorbike sound and great care is needed in the
design to ensure this is minimised without destroying valuable
character in the signal induced by the time varying frequency
component of $\beta(r,t)$.

From Fig.5 it is clear that the system bandwidth and the
rate of change of radiation frequency m are independent variables
which allow a choice of T_s to suit the human ear. This determines
the resolution limit, that is the theoretical minimum difference
frequency between two tone signals which can be unambiguously
detected. From Fig.7 this is seen to be $2/(T_s - \tau)$ because of the
line structure of the signal spectrum.

If an object consists of a number of surfaces each
scattering energy back to the receiver and the distance to each
surface is $r + \Delta r_i$ the audible signal corresponding to the object
will be

$$S_a(t) = \sum_{i=1}^{n} \beta(r + \Delta r_i, t) \sum_{p=1}^{\infty} a_{p_i} \cos p_i\, \omega_s t \qquad (9)$$

where ω_a in a_{p_i} is given by $2\pi \cdot \dfrac{2m}{c}(r + \Delta r_i)$

This narrow band of tones each representing a reflecting
elemental surface (like leaves on a tree) form a spectrum of
signals uniquely related to the geometry of the object relative
to the viewer. The sound therefore has a unique audible character

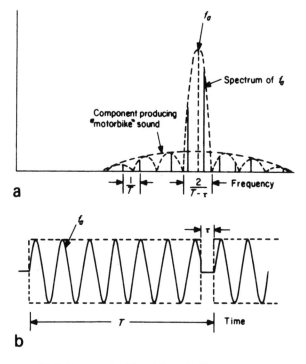

FIG.7.   Audio Signal Spectrum.

which provides a means for auditory object discrimination and even
sometimes recognition. It should be noticed that if, as in the
case of a tree with leaves fluttering in the breeze, the reflecting
elements are in motion we have to write $(r + \Delta r_i)(t)$ when each
element returns a Doppler shifted "replica" of the radiation
frequency and $f_a(t)$ is given by equation 7.

        There are thus several factors which uniquely determine the
signal representing an object creating the maximum information
transfer to the auditory system. It should be noted that the
Doppler shift in radiated frequency is unmodified in the audio
spectrum when a small fractional shift of the radiation frequency
may become a very significant shift in the audio band.

                Factors Affecting Choice of Distance Code

        The distance to the object of interest and the rate of
change of distance are represented in the audio frequency band by
equation 7. When there is no motion the audio signal is simply
$2mr/c$ and the variable m determines the frequency of f_a for any
given distance r. This is illustrated in figure 8 for two values

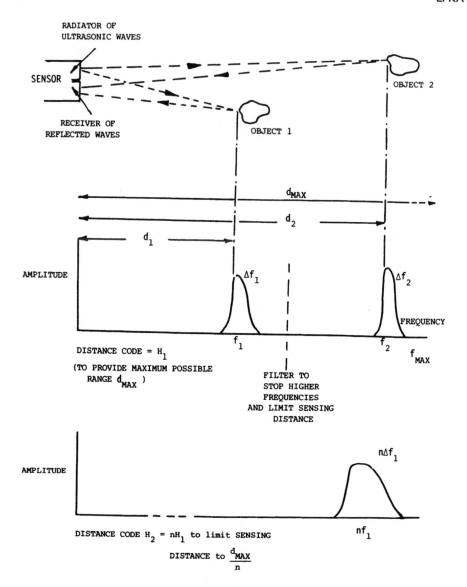

FIG.8.   Distance Code of Sensor.

of distance code H = 2m/c Hz/metre. The sensor as designed for
adult mobility on city streets uses a code H_1 of 900 Hz/m and the
maximum distance at which a wall can be sensed is arranged to be
approximately 6 metres producing an echo frequency of 5400 Hz.
When this distance code is changed to 10,800 Hz/m simply by
increasing the value of m, as is done in the aids used for babies,
the maximum distance which can be sensed is 0.5 m and the maximum
audible frequency thus remains 5400 Hz.

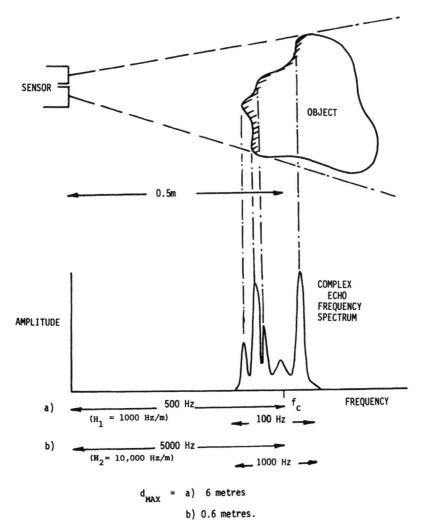

FIG.9.   Object Discrimination Using Two Extremes of the Distance Code.

The spread in frequencies Δf_a representing the object increases linearly as the distance code is increased and it might be assumed that greater discrimination between tone complexes would become possible as this spread in frequency Δf increased. This is so only to a limited extent as shown by Plomp and Steeneken (40); the ability to discriminate between sounds of differing quality is superior above 1000 Hz to that below 1000 Hz. From this work, as the mean frequency of the tone complex increases so also must the spread in frequency in order to retain the discrimination.   Beyond about 3000 Hz the spread in tonal frequencies must increase more in proportion to the rise in mean

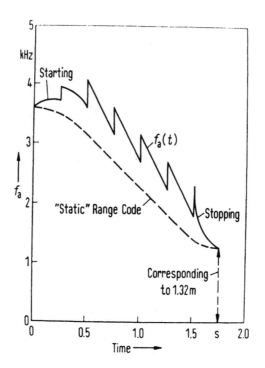

FIG.10.   Effect of User Motion Approaching an Object.

frequency. Consider the two codes shown in Fig.8 as used to
provide object discrimination at a distance of 0.5 m shown in
Fig.9. It will be evident that at these two extremes in the
usable distance code there will be a marked difference in object
discrimination. The long range code H_1 provides very poor
discrimination capability and the short range code H_2 may have
exceeded the optimum for an object at this distance. The actual
optimum has yet to be determined but using the variable range code
developed by Boys et al (10) it is now possible to "focus" the
sensor in distance so as to obtain the optimum for any distance
within wide limits.

When there is relative radial motion the audio signal $f_a(t)$
is determined both by the parameter m and the radiated frequency
$f_T(t)$ which is itself a time varying function. The effect of
walking up to a pole and stopping at a distance of 1.32 metres
(4 ft) is illustrated in Fig.10. The saw tooth in the distance
indication is due to the cyclic variation in $f_T(t)$. The error in
the indication caused by the Doppler shift is a significant fraction
of the instantaneous distance. It is explained later why this
does not cause any problem under user operating conditions.

TABLE 1

Sensory Aid Parameters

DESIGN

	A	B	C	D
Maximum sensing distance (r) in metres	2	1	1	0.5
Ultrasonic frequency band (kHz)	40–80	40–80	60–120	60–120
Sweep time T_s ms	96	48	72	36
Distance code H (kHz/m)	2.5	5	5	10
Maximum Doppler Shift (Hz) (maximum reaching velocity of 1 m/s)	500	500	750	750
Theoretical maximum resolution mm. (From Fig.7).	8.3	8.3	5.4	5.4

A more critical sensing situation is reaching for an object within arm's length where it is reasonable to expect that a blind child, for example, wearing a sensor would develop a degree of hand-ear coordination if it could sense the instantaneous position of its hand and the object of interest. A choice of four design parameters is shown in table 1 to illustrate the range of choice available to a designer of a sensory aid.

The auditory stimuli resulting from a reaching hand are shown in figs.11(a) and (b) where it will be seen that not only is there considerable stimuli variation but some of the stimuli are much less representative of the reach than others. The reaching indicator for designs A and C is very poor and for much of the reach the stimuli is inaudible (too low a frequency for the earphones to respond adequately). In designs B and D the indication of distance more nearly represents the reach: design D is thought to be the best. Quite evidently one design will not provide the optimum sensory stimuli for all situations and for a sensor to provide the best realisable spatial information it must have variable parameters under the control of the user.

Direction Code

The basic direction code is expressed in terms of the interaural amplitude difference (IAD) in decibels per degree

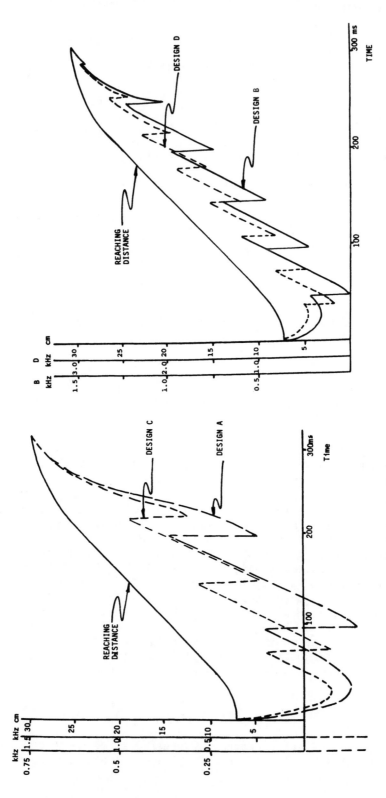

FIGS.11(a) and (b). Audio Indication of a Reaching Hand for Four Different Parameter Sets.

azimuth angle and can be written in terms of the well known
relationship for the direction of a low frequency tonal image
reported by Stewart and Hovda (47). Thus the estimated angle of
an object is given by

$$\theta_E = 20 \ K \ \log \frac{A_L}{A_R} \tag{10}$$

where A_L and A_R are the sound levels of the stimulus at the
left and right ears respectively and K is the subject's constant
of lateralization of binaurally presented sounds. Ideally one
would expect the estimated angle θ_E arising from the lateralizing
sensation from the sensory aid should be the same as the actual
angle, or a known function of it. This is not easily arranged for
each individual aid user because the lateralizing constant K varies
between individuals and each one has to undergo special tests to
determine it (42,36). Even if the IAD could be readily adjusted to
suit individuals, the technical problems are such that it may be
suitable for only a portion of the distance being sensed. This is
partly because the lateralizing constant for each individual varies
with frequency, and partly because the technology for obtaining the
direction code has inherent physical limitations. An engineering
compromise is necessary, although it is not yet determined what this
should be.

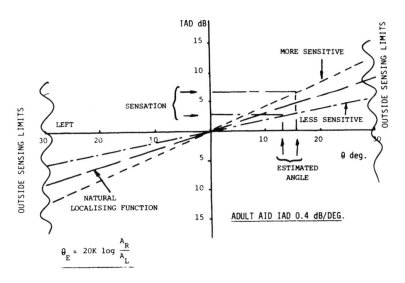

FIG.12.   Ideal Direction Code for Different Sensitivities of the
           Localising Function.

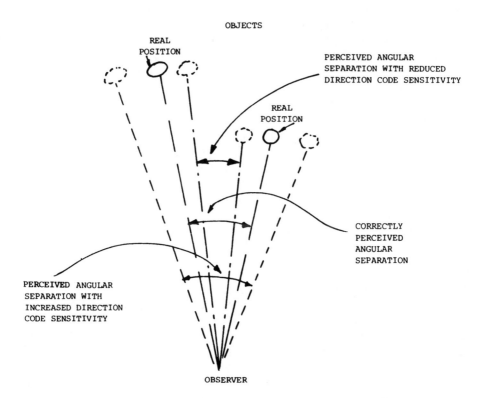

FIG.13.   Effect of Varying the Direction Code.

An important aspect of the direction code is the freedom to choose the degree of angular resolution built into the sensory aid. The direction code is shown in Fig.12 where it will be seen that it can be set to have increased or decreased sensitivity relative to that which corresponds to the natural lateralizing function. In Fig.13 is shown the spatial effect of changing the direction code. When the code is made more sensitive, objects close together in azimuth angle may be perceived as being separate when otherwise they may have been thought to be in the same direction. A very close analogy may be drawn with changing optical lenses so as to expand the visual image in order that more detail can be seen.

There are technical difficulties in obtaining the ideal direction code given by equation 8. To obtain this the azimuthal field response of the two receivers as measured at the audio output was shown by Rowell to be (42)

Left receiver $A_L(\theta) = C_L \exp [-(\theta + \alpha/2)^2]/2\alpha K$ (11)

Right receiver $A_R\theta = C_R \exp [-(\theta - \alpha/2)^2]/2\alpha K$ (12)

where C_L, C_R are constants and normally $C_L = C_R$

    α is the angle of divergence between the polar axes of the two receivers.

Thus the polar plots follow a Gaussian function displaced by an angle $\alpha/2$ to the left and right of the median $\theta = 0°$ so as to create the interaural amplitude difference direction code in degrees/dB. To obtain the desired distance code a broadband radiation of the order of one octave is necessary. Thus the wavelength may change by a factor of 2:1 during the frequency sweep. It follows that without special time varying aperture shading the polar response of the ultrasonic transducers must also vary by a factor of approximately 2:1. The value of K in equations (11) and (12) is the desired localisation constant to match the user, but it will be seen that it is now time varying over the sweep period causing the estimated angle θ_E to vary as A_R and A_L vary for any given angle θ. It would seem that such an arrangement could not provide the desired direction cue. It was discovered however that during the sweep period T_s the auditory system did not respond to the instantaneous ratio A_L/A_R but to some integrated value corresponding to the energy difference in the period T_s.

Using solid dielectric transducers described in (26) the desired Gaussian response could be obtained to an adequate approximation over the field of view provided

$$\theta_E(\text{measured}) = 20 \text{ K Log} \frac{\int_{t=t_n}^{t_{n+1}} A_L(t)dt}{\int_{t=t_n}^{t_{n+1}} A_R(t)dt}$$ (13)

It is here that the greatest technical difficulties lie in making the sensor. An approximately linear relationship between θ_E IAD (dB) can be obtained over the field of view provided this is restricted to an angle 3α by the radiation field of the transmitting transducer. Direction ambiguity will otherwise result as shown in Fig.14.

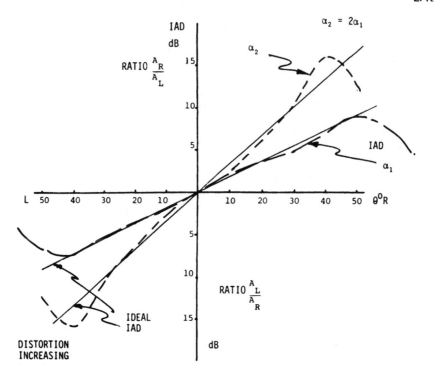

FIG.14.   Distortion in Direction Code Due to Physical Limitations
          of the System.

     The binaural direction cue thus relies entirely upon the
lateralizing sensation produced by the dichotic presentation of tone
complexes. It cannot therefore immediately create an "out there"
or spatial sensation in a naive listener; only the equivalent of a
pina in the echo signal path can do this. Nevertheless, long term
users do say they experience an "out there" sensation which is
probably induced by user motion.

Field of View

     Probably the most difficult of the several sensory parameters
to appreciate without personal experience is the field of view
through which objects may be perceived. Imposed upon the cues to
distance and direction is the variable of object reflectivity which
determines the strength of the echo and the variation of echo
intensity caused by the change in system sensitivity with direction
and radiation frequency. The dynamic range of the ear allows
perception of objects well outside the field of view expressed as
a cone of solid angle Ω representing the half power sensitivity and
greater. Whilst this is a technical convenience and commonly used

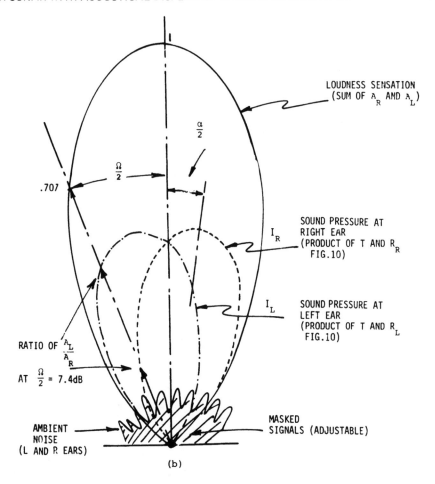

FIG.15.   Illustration of Field of View of Binaural Sensor.

in sonars with visual displays it is very misleading in an air
sonar with auditory display.

An illustrative plot of a typical field of the Binaural
Sensory Aid is given in Fig.15 and derives from the transducer
sensitivity plots illustrated in Fig.16. (It should be
remembered that the transducer plots can be for only one frequency;
the field plot is taken to be the integrated audio output). The
"loudness" plot is entirely illustrative.

In the vertical plane the field sensitivity varies in a
complex way because of the splay angle α and has a significant
effect on the I.A.D. characteristic for angles off the horizontal.

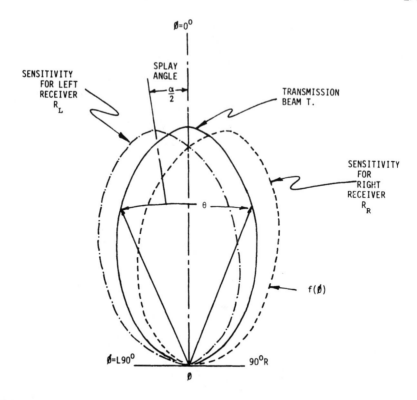

FIG.16.   Typical Transmitter and Receiver Field Plots in Azimuth.

        The user of an aid does not hear the left and right sounds
separately, but instead it is a fused signal which shifts to the
left or right according to the I.A.D.   As the head is turned
about the direction of an object, if the I.A.D. matches the
lateralization function, the real direction and the sensed
direction will coincide giving the impression that the object
remains still.

        However, because the instantaneous sensitivity of the field
of view varies with frequency "off axis" in a different way to
"on axis", the character of the object echo also varies as the head
is turned. Object recognition is thus more reliable when "looking"
at the object than when "looking away" from it.

        Varying the splay angle has only a secondary effect on the
the field of view as plotted because of the radiation response but
it has a marked effect on the detection of an object entering the
field from the left or right when another object is already in the
field. This is because with a large splay angle one ear hears the
echo more distinctly, thus increasing direction discrimination.
It is thus evident that whilst the distance code is independent of

the field of view the direction code and field of view are
dependent variables and must be jointly considered in any design.

OBJECT DISCRIMINATION AND MOTION PERCEPTION

Distinguishing one object from another when more than one
object is present in the field of view requires that two or more
objects can be separately resolved. This is so whether perception
of the objects is by visual or auditory means. When the objects
are widely spaced in direction so that only one can be in the field
of a sensory aid, all that is necessary so as to determine their
relative positions is turn the head. The viewer then seeks to
recognise one object from another by correlating the echo signal
with previous experience in sensing objects. This was tested by
Sharpe when he carried out an evaluation of the "Kay" Sonic Torch
in 1968 (46). It was established that, using the wide band CTFM
air sonar described in reference 25, subjects were able to
recognise a variety of objects and surface structures including
different trees, various forms of fencing and objects forming
landmarks. The study was based upon the training procedure designed
by Elliot et al (19). A special series of tape recordings provided
the basic training material prior to experience in the real world.

The binaural sensor in the form of spectacles or headband poses
a problem in defining resolution from which the degree of
discrimination can be determined. It was shown earlier that the
maximum theoretical resolution is given by $2(T_s - \tau)$Hz from which
the spatial resolution may be determined from $2/H(T_s - \tau)$ metres
where H is the range code. It was shown that for design D (Table 1)
this is 5.4 mm. In terms of auditory resolution it implies that
two echo tones representing two vertical fine strings at say 250 mm
in the field of view can be perceived as distinctly different from
one string at this distance. This is so; but only because the
sound is different, not because the two strings can be perceived as
being separate - as is the case for optics. In order that the
echos can be heard as distinctly separate the strings need to be
separate in distance by 100 mm when the farthest one is at 250 mm.
This is discussed by Kay and Do in references (27, 18) and by Do in
the next paper of this chapter. For less separation the two echo
tones become combination tones as described by Fletcher (20) and
confusion is experienced by subjects under test when questioned
about what they hear (42). In an attempt to describe the spatial
resolution capability of a user of the sensory aid as distinct from
determining the ability to discriminate between or recognise objects
Do and Kay (18) simulated the sensory aid auditory space using an
analogue computer to control the frequency of two or more tone
generators. One object in the field of view was simulated as moving
relative to another either in the same direction (same IAD) or in

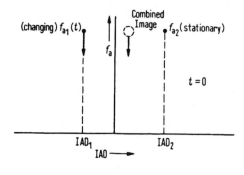

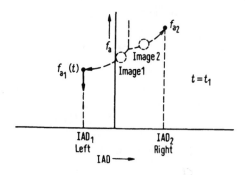

FIG.17.   Simulated Moving and Stationary Objects to Test
Auditory Resolution.

different directions (differing IAD). A simple simulation is
illustrated in Fig.17. $f_{a_1}(t)$ represents an object moving towards
the viewer. When the two objects are at the same distance the
combined auditory image lies midway between the two images if
perceived separately. As the changing frequency $f_{a_1}(t)$ decreases
the sound changes in character from a tone to increasing beats and
the image glides down in frequency until at a critical frequency
difference it must split into its two separate components. This
follows because at a difference frequency of 40% of the higher
frequency the two tones are perceived as separate even under
stationary conditions. It was discovered that the separation may
occur for as little difference as 10% of the initial frequency.

Do in the accompanying paper shows that for tone complexes this difference can be only 5%, almost 10 times better than the stationary state.

This has considerable relevance to a spatial sensor for the blind because motion of some kind is constantly taking place. Turning of the head also induces frequency glides which may influence the spatial resolution but this requires detailed study (17).

There are thus three forms of spatial perception determining how well (or poorly) a blind person can view his immediate environment.

a) Identification of an object by virtue of the sound character it produces determined in the limit by the theoretical resolution of tones.

b) Discrimination between objects due to the difference in sound from each which can be detected through sequential comparison.

c) Resolution between objects either stationary or in motion by virtue of the ability to listen to each separately even when they are simultaneously producing sounds.

These three forms have not been studied adequately but raise interesting theoretical issues as discussed by Do. Until they are well understood it will not be possible to determine the optimum design of aid.

BEHAVIOURAL STUDIES WITH THE BINAURAL SENSORY AID

A sonar design can be tested by well established measurement techniques to ensure that it meets the required engineering specifications. Evaluation of the sonar on the other hand in terms of usability - which seems to be the only valid criterion for a sensory aid to spatial perception - involves the total life function of the particular individual who will wear it. Only under day to day living conditions normally experienced by a blind person can an aid be assessed to see if it has user value. Generalisations about usability from an adequate sample of the user population may then be possible. Laboratory type experiments can be used to determine the level of attainment of a particular skill which may be acquired with an aid. There is however no way to determine the value of that skill to the user of an aid until many other skills have also been acquired and incorporated into a modified way of living. Those skills of little value then lapse into disuse. It is then that newly acquired skills of significant value can be tested. This process poses a

dilemma for both the designer of an aid, who is given no guidance
in the setting of system performance specifications which could be
inappropriate, and the agencies serving the user population who have
to determine the cost-effectiveness of their service in possibly
providing an aid. Two special conferences (39, 22) on this matter
were unable to resolve the many differences of opinion which existed
in the early 70's and indeed continue to exist today, if perhaps to
a lesser extent. A study undertaken by behavioural scientists to
determine the value of a sensor under controlled conditions
divorced from both the user's day to day emotional problems and the
designer's professional interest was thought to be the most
appropriate approach by several consultants: in the end, however,
general acceptance of the Binaural Sensory Aid as described by
Kay (26) came about through the hard school of user, professional
worker and designer interaction during an extensive evaluation
programme in the period 1970-72 (26, 28).

                          Adult Sensory Aid Use

     To be of value a sensory aid to spatial perception should
provide the user with a useful percept of the immediate environment
over an adequate field of view. Since the Binaural Sensory Aid
output is sound, this can provide only a little of the information
sought by a blind person. On the other hand, the environment in
which we travel is so structured by human design that vision is not
essential in order that individuals may identify many features of
their environment. For example, many sound producing objects are
readily identified by their sounds alone although this should
not be construed as implying the stimulation of a mental image.
In a similar way, objects causing echo sounds in the sensory aid
can also be recognised by their sound character provided the
character is reliably repeatable and there is sufficient a-priori
information available for the recall of experience. For example,
walking along the street in a shopping area it is common to find
the pathway lined with parking meters. Blind people would normally
be unaware of their presence but when wearing the binaural sensor
it would produce a regular series of sound patterns every few steps
which seem to come from the side of the roadway.   These sounds can
quickly come to mean "parking meter". What this image constitutes
in the congenitally blind remains unknown.

     In evaluating a sensory aid this kind of perception must be
carefully defined and the cognitive process employed by the aid
user considered in depth otherwise the wrong goals may be sought.
When a sensory aid only indicates the presence of a single object
at a time, as in the case of "clear path indicators", the user
information is trivial and evaluation relatively simple. Even so,
this information can reduce stress in the Long Cane traveller who

relies greatly on his natural hearing for oeientation. Evaluation
of the binaural sonar, instead, highlighted the need for a general
theory of mobility and an international seminar on the subject was
held at Oxford in 1973. The only paper to reach publication from
this seminar was that by Kay (29) expressing some thoughts on a
theory arising from the extensive observation of blind travellers
during the binaural sensor evaluation. This has now been carried
further by Brabyn (13). During the evaluation process some 25
Orientation and Mobility specialists whose primary role is that of
teaching mobility to the Blind were themselves trained to use the
sensor under blindfold over a period of four weeks. Although this
was thought to be an inadequate period of training, it was the
longest time likely to be made available and was planned accordingly.
These specialists then, between them, trained about 100 blind
persons, who had either Long Cane or Guide Dog travel skills, to
use the sensor in a variety of travel situations including downtown
travel. The pattern of training was built round the Long Cane
training programme so as to integrate the two aids to mobility.
In consequence, the binaural aid became a secondary aid to mobility
(the Long Cane being the primary aid) and its role as a spatial
sensor was largely suppressed in favour of obstacle detection and
"clear path indication".

On completion of the evaluation programme the 25 teachers and
96 users were sent confidential questionnaires by a professional
field service evaluator who designed the instrument (1). A number
of cross correlations were carried out from which it was found that
the responses from users cross correlated better than those from
the teachers, and there was often poor correlation between teacher
and user response. In Table II, for example, (from Kay, 30), the
responses from the Long Cane user and Guide Dog user are separated -
showing quite different responses to certain questions. It was to
be expected that in some situations a guide dog would provide more
assistance than could the information from a sensor with limited
distance sensing capability. Hence, from the question on "Crossing
a street in a downtown area" the guide dog user gained little
benefit from the aid since the guide dog is trained to give
assistance. The long cane user on the other hand did benefit as
is evident from the fact that 68% of the users reported improved
mobility. The teachers thought the aid had little impact on
mobility in this situation. Subsequent questioning of groups of
users revealed that the teachers had not taught the use of the aid
in this situation and therefore were making inexperienced judgments
compared with users who had tried the aid in this situation
subsequent to training. Little detailed analysis of the
questionnaire is available in the literature. Airasian discusses
a number of factors which may account for the discrepancies in the
72 user responses finally received and the 21 teacher responses:

TABLE II

The Sonic Glasses: Ratings of Usefulness in Mobility.

Skills	Percentage of Users *				Percentage of Trainers (N = 21)		
	Greater Mobility	Less Mobility	No Change	Never Encountered	Great Impact	Mod-erate Impact	Small Impact
Locating landmarks	93/100	2/0	5/0	0/0	58	42	0
Avoiding pedestrians in crowded areas	73/18	5/18	20/64	2/0	30	65	5
Avoiding obstacles in one's path	93/36	2/0	5/64	0/0	60	40	0
Following sidewalks bordered by grass	29/9	10/0	56/91	5/0	0	0	100
Locating doorways	80/70	2/0	17/30	0/0	58	37	5
Guidelining (shorelining)	80/36	5/0	13/64	2/0	37	42	21
Following paths with hedgerows	88/36	5/0	7/55	0/9	37	53	10
Following paths in hallways	61/27	12/18	17/55	10/0	35	55	10
Locating a clerk in a store	29/45	0/10	32/45	39/0	30	35	35
Taking one's place in a line	49/64	0/9	17/18	34/9	45	40	15
Locating a seat in a restaurant	17/9	7/0	17/64	59/27	15	40	45
Crossing a street in a downtown area	68/18	7/0	20/82	5/0	10	15	75
Traveling in moderate pedestrian traffic	76/45	2/9	20/45	2/0	21	63	16
Finding a specific store downtown	34/55	7/0	32/45	27/0	25	60	15
Seeking pedestrian assistance	46/30	5/0	32/60	17/10	40	30	30
Finding one's way in unfamiliar suburban area	59/55	7/0	12/36	22/9	15	55	30
Finding one's way in familiar suburban areas	71/36	2/0	22/55	2/9	20	60	20
Finding one's way in familiar shopping centers	71/45	2/0	12/45	15/10	60	25	18
Finding one's way in unfamiliar shopping centers	56/36	5/9	7/18	29/36	20	55	25
Finding one's way in familiar rural areas	30/40	7/0	8/30	55/30	26	42	32
Finding one's way in unfamiliar rural areas	27/50	7/0	2/20	63/30	11	47	42
Finding up curbs	38/27	10/0	50/73	2/0	0	10	90
Assessing the distance of stationary objects	98/82	0/18	2/0	0/0	70	30	0
Assessing the distance of moving objects	56/40	10/10	32/50	2/0	15	70	15
Squaring off to face objects or people	73/45	2/0	17/55	7/0	55	35	10
Finding one's way in large open spaces	34/9	2/0	49/91	9/0	15	5	80

*First figure represents percentage long cane users (N=43); second figure percentage dog guide users (N=11).

one of them relates simply to the basic social problem of
"diffusion of innovation".

In an attempt to obtain objective analysable data on sensory
aided skills, there was built into the "standard" training programme
a series of well defined exercises in the form of Lessons which
students were expected to complete before moving into everyday
travel situations. These were to provide the means to record
performance such as time to achieve specified levels of achievement.
An example of an exercise is shown in Fig.18. Further details may
be found in reference (31). The blind student was required to walk
centrally in a straight line between two rows of poles then slalom
back along one row. In terms of body control the task is not
trivial. Almost no records were kept by the teachers. However
five teachers under training were filmed executing this task and
the film was subsequently shown to 58 "judges" (a class of university
students) who were asked to rate the performance of each teacher on
a scale from 0 representing "blind" to 10 for "sighted" performance
(29). Their judgments are shown in Fig.19. There is considerable
variability in these showing how unreliable any individual judgment
may be for even a simple task. How much more variable then are

FIG.18.   Typical Pole Exercise for Testing Locomotion Skills
            Using Sensory Aids.

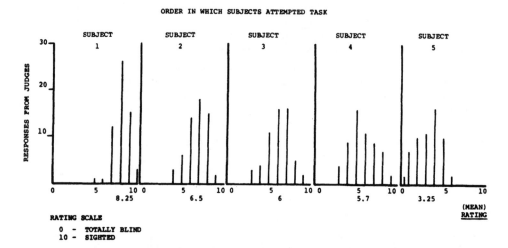

FIG.19.   Response of Judges After Viewing Film of Five Subjects
           Carrying Out Exercise in Fig.18.

judgments of mobility likely to be for real life situations and
the extent to which a sensory aid plays a part is exceedingly
difficult to determine.

    The skill shown by one blind subject in executing the slalom
is illustrated in Fig.20 which was produced from an anlaysis of
film of the action. Each footstep is plotted together with head
direction shown by an arrow. Brabyn,et.al.(12) reports an attempt
to produce a more objective measure still, by using a tracking
system in a specially treated room 20m x 13m. It was quickly
realised here that a meaningful definition of the ideal path and
deviation from this in order to find a figure of merit was difficult
to obtain except for walking a straight path at a specified
distance from a guideline - row of poles. For more complex tasks
such as the slalom a track of a subject's path was easily obtained
but an "ideal path" is less easily defined: a variation in path
between subjects did not mean a variation in the skill of executing
the task.

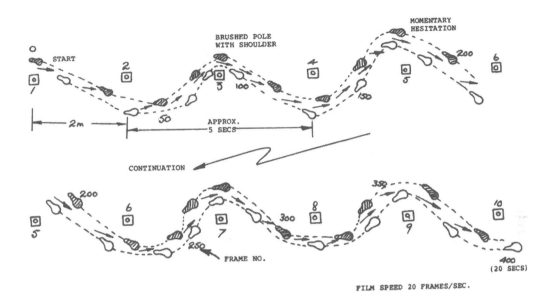

FIG.20.   Blind Subject Slaloming Round 10 Poles.

It became evident that accurate measurement of sensory aided
mobility was easiest for straight path travel where it was found
that blind sensory aided subjects could perform as well as sighted
in terms of the measurements taken. Variation in the direction cue
of the sensory aid had considerable effect on the subject's
performance, but changes in the distance cue had only small effect.
This is thought to be due to the generation of auditory flow
patterns which are more affected by the direction cue than the
distance cue (13). The detection threshold for a change in the
direction cue is important in determining a subject's path. When
the threshold is small, error correction can be made even during a
step as one foot is moving forward. It seems it is the auditory
flow pattern which becomes highly developed with sensory aid use,
that provides the primary guiding information for smooth locomotion
in a cluttered environment. The experiments of Do certainly show
that this kind of motion generates the greatest degree of object
resolution. It was observed that the visual looming effect of
Gibson (21) seems to apply equally well to spatial auditory
perception through an acoustic spatial sensor. It will be realised
that blind sensory aided performance approaching that of sighted

performance on one task does not lead to the general conclusion that the Blind can now travel like the sighted. Such is the evaluation problem.

Film records of totally blind persons travelling in a busy downtown area using a spatial sensor do nevertheless show that "near" sighted performance by a few is possible as judged by the blending of the blind person into the pedestrian flow. Thornton (52,53) describes his experience in using the sensor during a period of extensive travel on foot but the individual personality characteristics which lead to this proficiency have not been determined. They are not restricted to an intellectually elite group although such a group has more need of a spatial sensor in their travels than do some others. Much too little is known about the human characteristics which lead to good independent mobility among the blind, and still, inadequate information is available from which to design the optimum form of this aid. Two decades of research have so far failed to quantify the problem sufficiently; it is so complex.

Use of a Sensor by Children

The use of a spatial sensor which provides an entirely new form of perception of the environment was often said to be more suitable for children than adults. It was thought the children would learn the spatial cues more quickly than adults who were said to be slow in learning complex sound patterns and they could be more interested in their surroundings. Bower (8) makes this kind of point very strongly in his argument for using a sensor on a baby although he has not produced any evidence in support of his theory. Contrary evidence has now been obtained by other experimenters (48), yet at the same time there does seem to be reason to expect that babies and small children may come to adapt more completely to a new sensor than older children and adults who may already have established strong habits in their behaviour. A well developed reliance on tactile information for example may inhibit later adaption to auditory information. It was this kind of argument which led to the study of sensory aid usage by children at Canterbury (49). The acceptance of the Binaural Sensory Aid for adults in 1972 following the evaluation opened the way to use by children, but the sensory parameters used by adults were unsuitable. They had been chosen for mobility out of doors in conjunction with the long cane. Children have different needs, discussed by Strelow and Hodgson (49) which called for considerable changes in the aid design following a period of observation of blind children (48).

The choice of sensing distance is of prime importance for it is a fundamental factor which must influence greatly the rate of learning by a child. A long range sensor collects large amounts of

information requiring a high degree of object discrimination if the
many objects which can be simultaneously sensed are to be perceived
as separate entities. The binaural sensor, particularly with a
medium to long range, does not provide this discrimination in a
stationary environment as perceived by a baby. This has already
been shown to be so by Do. Close range objects which produce low
frequency sounds are also easily masked by the dominant low
frequency components of ambient noise contaminating our environment
(44). They are therefore less reliably perceived. For a developing
child these close objects represent the most important part of its
physical world.   On the other hand, as discussed by Warren (56),
a child which has reached some understanding of its immediate
surroundings and is ready to develop locomotor skills can be
expected to be more motivated to explore distant surroundings if it
can sense that these exist. It would seem reasonable then to
provide an aid which, initially, could be adjusted to provide
greatest detail about the close environment - which can be kept
uncluttered - and then later increase the sensing distance as the
need or interest develops in the child. With this approach the
complex effects of motion on the distance indicator can be most
readily accommodated.

     An aid design evolved which provides a variable maximum sensing
distance and an automatic level controller which maintains an
almost constant signal to ambient noise ratio for an ambient noise
change in excess of 20 dB. It is arranged that the user is unaware
of the control taking place (22). It is now feasible to arrange
for the variable sensing distance to be under the control of
muscle potentials. The field of view may be chosen during the
design stage of an aid, but at present, once built it remains fixed.
This limits the research which can be carried out because the field
of view is an important parameter in object sensing and
discrimination. A narrow field of view of say 5°-10° provides
remarkably good auditory discrimination in azimuth but only radial
motion along the axis of the field may be perceived. Head
searching, no matter how slight this may be, is necessary to
determine the relative position of objects.

     A wide field provides good motion perception but azimuthal
discrimination is poorer due to the multiple tone complexes
produced by several objects. Both fields have special value to a
child learning about its world but at present both are not
simultaneously available. The requirement may be likened to
central and peripheral perception, but in an auditory sense. It
now seems feasible to create this artificially.

     A small number of blind children have been exposed to sensory
aided perception and it is becoming evident that blind children in
general who are about to become mobile or who have become mobile
may learn to perceive their environment more effectively through

the sensor. The extent to which they do this must vary and is undetermined at this time. The kind of experiments carried out with animals to establish their use of a natural sonar cannot be undertaken with children. Instead, use of the artificial sonar forms part of a complex life style - even for a small child - and development on a wide front is constantly taking place. Strelow, et al. (48) discusses some operant training which was used but explains why this lasted for only a few sessions before the research procedure developed into case studies on development of two children which is partly reproduced here.

Anophthalmic Boy. A totally blind 2½ year-old boy, whose physical development was good, was provided with an experimental aid to determine if the increased sensory stimulus would bring about a change in his developmental pattern. The aid parameters were the most appropriate that could be provided at that time.

While no other handicaps were assessed, he showed a serious developmental delay both in walking and in acquisition of language. He showed no form of walking or crawling on his own, although he could walk when supported by parents and pull himself into a standing position against a wall or piece of furniture. His language was restricted to babbling and cooing sounds. Social interaction was nonexistent; he showed no obvious awareness of people around him unless touched or tickled, and did not reach out for objects making sounds such as rattles and bells. He was not even startled by loud noises.

It was decided to see whether he could be trained in a spatial activity by teaching him to reach for the cup when it was brought in front of him and sensed only through the aid. A small amount of milk in the cup served as reinforcement. By training the boy to reach to the sound cue, relating to his drinking cup, it was hoped to teach him to attend to the device sounds and to relate these to objects in space. The training task had elements of both classical and operant conditioning. The response of reaching to the sound of the cup being tapped, was transferred to the signal of the cup, sensed through the aid, in a manner analogous to classical conditioning. The reaching response was "shaped" to its final form in a manner analogous to operant procedures.

The reaching study established that a spatial task involving the aid could be learned by the boy. However, the extent to which the boy perceived the aid sounds as spatial information meaning, "An object is out there" as opposed to merely reaching on command to sounds in the ear was not established by this comparatively simple training procedure. More elaborate procedures in which objects were placed in different spatial positions with the boy required to respond differentially could have established this point.

It was concluded by Strelow et al. after a considerable period
of training, that a developmentally delayed child who was initially
immobile and unresponsive to auditory stimuli, showed a major
improvement in locomotion skills and a marked improvement in
listening skills. While some of this was attributable to the
intervention activities per se, and some due to normal developmental
progress, the sensory aid appeared to help at critical stages of
development.

Congenitally Blind Girl. A 6½ year-old, congenitally blind
girl was in many ways a complete contrast to the boy. She was a
verbally precocious child of well above average intelligence (IQ
125-135), who attended an integrated school that had a visual
resource center. A month's regular observation indicated virtually
no independent locomotion. Requests for her to walk on her own
were met with stubborn refusal, a form of behaviour that was
particularly common when she was observed at home.

At first it was impossible to obtain cooperation for periods
longer than 2 or 3 minutes, but within a few sessions sustained
work for periods of 10-15 minutes was possible. The initial aim
of the work was to establish the extent of her ability to use the
aid in the Sensory Perception laboratory. A series of tasks was
tried in an effort to develop some skill in discrimination of aid
sounds.

Chasing the experimenter - The girl attempted to follow the
experimenter and was occasionally allowed to catch up. This was
felt to be particularly useful for encouraging quick response to
the spatial information provided by the device. The large Sensory
Perception laboratory was sound-deadened and carpeted, and ambient
cues were almost nonexistent.

Shoreling - The task required walking at arm's length along a
row of six poles spread one meter apart. At first she was allowed
to touch each pole as she went, but was later required to maintain
her distance by sound alone and walk up to the last pole.

Slaloming - This was performed with the same pole arrangement
as in shorelining.

Beacon Exercises - An omnidirectional ultrasonic radiator,
which could be detected only through the aid, was used as a homing
beacon. The direction could be sensed binaurally and an indication
of distance was gained from the strength of the signal. The device
was carried by the experimenter to provide a cue to his position
in games of hide and seek, even when the experimenter was hidden
among objects.

This regime was interspersed with occasional walks around the campus, holding hands, during which objects of interest were pointed out. The laboratory exercises were maintained until some level of skill was noted. The most successful task for maintaining interest was the chasing task. The pole exercises were at first the least interesting, although also the most demanding in requiring precise control of body motion in relation to objects.

The first travel task attempted was a walk between the girl's classroom and the visual resource center. There was an irregular concrete travel path between the two buildings requiring a number of 90° turns. At the beginning of the path there was grass on one side and buildings on the other, but it gradually opened up. The most difficult obstacles for an unaided blind traveller were metal poles of an outdoor gymnasium, although they were easily detected with the aid. A final open area led to a large tree in front of the resource center, a total travel distance of about 150m. There were no reliable ambient orientation cues and there were irregular gaps between the buildings so that a simple shorelining procedure was not possible. Some form of navigation skill had to be developed, and the child taught to recognise important objects. Since the school was for sighted children, little attention was given to keeping the travel routes free of unexpected obstacles, so the girl had to be able to cope with them.

The girl's fear of solo travel was at first hard to combat, but by walking with her over the route, taking her by the hand, she began to develop the confidence to move on her own. The experimenters travelled with her for about two weeks, monitoring her aid sounds and explaining what they signified. Finally, she was tested in her ability to handle the route entirely on her own, which she did, albeit slowly and awkwardly, but nevertheless safely.

It became important at this point to ensure that the girl was competent to look after the aid on her own, able to change batteries and bring it to school each day. To earn permission to keep the device, she had to pass a test, which included handling of the aid and batteries as well as performance on shoreline and slalom in the laboratory. The girl was surprisingly serious about the two practice sessions and the test itself. Her performance, while not outstanding, at least indicated a level of competence; two of the five slalom tasks were judged as good, as were three of the five shoreline trials. She was then allowed to keep the aid at home and bring it in herself each day.

To go to the visual center, it was necessary for the girl to put the aid on in class, pack up her work material, and proceed to the visual center on her own as required by the timetable. With the cooperation of teachers this soon became an established routine no longer worthy of comment by either teachers or the girl. Her performance was monitored daily by the research team but this soon

became an unnecessary precaution. She began to have a more
confident gait, and at no time was she observed to collide with or
trip over objects while using the aid. When occasionally she
wandered off course, she kept going until the next object was
sensed, walked up to feel it, and re-oriented herself.

Having established that the girl could use the device outdoors
as a primary aid, other settings were sought. Free use of the aid,
during recess periods and lunchtime, was not encouraged for a number
of reasons, primarily because the aid lacked an automatic level
control to cope with the very noisy environment during playtime.
The travel route to the visual center was taken during a quiet time
in the day when few children were about. However, it was decided
to attempt a more difficult travel route, which would allow her to
come into her classroom from the other side of the large schoolyard.

She arrived each day by taxi, which left her at the gate of
the schoolyard from which she was escorted to class by a teacher or
another student. The school environment was comparatively quiet at
these times (by comparison with the break periods), and a route was
charted out that allowed shorelining of fences and buildings. The
route was lengthy, required a good deal of turning, and there were
frequent unexpected obstacles. Considerable patience was required
by the experimenters and strong support from the teachers was again
helpful in convincing the girl that she could handle the task and
would soon be expected to do it on her own.

Again, daily training was given for two weeks, during which she
was brought to school by one of the team and accompanied on the
route. Her aid sounds were monitored by headphone so that objects
could be pointed out, and various travel strategies could be
explained. The time to travel the route dropped from 30 minutes
to about 9 minutes.

When she was able to perform the route successfully, she was
observed for another two weeks to make sure that she was coping
with the route. The precaution turned out to be unnecessary, as
her performance over the route soon became very good. Her gait
loosened up well and protective actions with her hands became
minimal. However, when unexpected obstacles were encountered, she
would walk up to them and place her hands very accurately to find
out what they were. She travelled quite fast on sections of the
route and generally ran the last part to the classroom, hitching
her school bags up under her arm and appearing little different
from a sighted child.

General Comment. It was discovered during the early stages of
training children in the use of the sensory aid that they learned
the apparently simple tasks less quickly than adults. The pole
tasks used in teaching adult mobility were eventually mastered and

performed proficiently by a very intelligent child but the time
scale of learning greatly exceeded that for adults. Outdoor
mobility in the playground of a school for sighted children also
came slowly but reached a higher level of attainment without the
additional aid of a cane than that seen for most adults. The child
occasionally broke into a jog even when there were other children
around. No blind adult did this during the reported studies.
Baird, Newcommer and Scionie (5,38,45) also report on sensory aid
use by children during lessons but these children do not get to
keep the aid and it is shared by the users. Little information is
provided on the levels of aid use each child reached.

Babies

The controversy over the use of a sensory aid by babies
resulting from a case study by Bower involving a blind baby raises
some fundamental questions (8,9,32). The binaural sensory aid has
been shown here to code the environmental information into an
acoustic form previously unknown to man: even the direction cue is
artificial in that the initial sensation is one of lateralization.
Hence the baby reported by Bower to have tracked an object with
its blind eyes through the sensory aid stimulation could also have
been induced to produce this response using the simple presentation
of dichotic sounds from standard earphones. Any lateral shift of
the fused auditory image should equally well have caused a
corresponding rotation of the eyes. Taking this further, the baby,
without prior knowledge, would have to attribute to the lateral
auditory shift an appropriate angular shift of the eyes so that it
appeared to look at the object represented by the sound. Film of
the baby as viewed by the writer was unconvincing. It could not
be established that the IAD direction cue was matched to the baby's
lateralization function and it is suggested here that any apparent
tracking of the object by the blind baby which was observed could
only have been fortuitous. Bower also argued that "placing" of
hands by the baby was sensory aid induced. This is quite feasible,
particularly since the auditory sensation when approaching a close
object is not unlike the visual looming discussed by Gibson (21).
This visual looming effect is very noticeable when viewing an
approaching light through frosted glass. Unfortunately, Bower in
his tests used two spaced specular reflecting batons as objects
and these were equidistant from the baby. As shown earlier in Do's
experiments and it is discussed elsewhere (34), two echos from
different directions but equal distances produce a single ghost
echo from midway between the objects. Hence, if the baby in a
placing position were stimulated to outstretch its hands by the
aid, it could only be the ghost which did this. How then did the
baby work out that there were two objects on either side of the
ghost and reach towards them as explained by the experimenter?
This, without turning its head to the extent that is necessary to

cause object separation (17). These apparent facts fit a theory proposed by Bower but the physics of aid performance do not fit the same facts. It is essential when experimenting with an aid such as this, that its physical performance be well understood and more should not be inferred from behaviour patterns than is physically reasonable to expect. What has to be realised is the difference between sensing the environment with light waves visually and acoustic waves audibly. The 1 to 10,000 ratio in wavelength itself processes out almost all the spatial information we are familiar with through the visual channel. Perhaps it is because some adult blind people have been able to demonstrate quite remarkable skills using the aid that some experimenters have come to believe much more was possible simply by exposing a very young receptive brain to the spatial information.

A wavelength of the order of 2.5 mm does in fact have the effect of creating mirrors in the environment because of the high proportion of object surfaces which are specular reflectors at this wavelength. Loudness of echo rather than structure becomes a dominant cue in many situations so care is needed in the design of any experiment. Bower, for example, used a cube suspended on a string as an object. This is highly specular as a reflector and as it freely turns, causes great changes in echo strength but, more importantly, it can cause strong left-right auditory sensations as the acoustic highlight of the radiator is reflected more to the left receiver than the right and then back again. A more appropriate object would be a sphere with a rough surface or, if large enough, a smooth sphere.

A number of babies are presently learning to perceive their environment through the medium of an aid: not all the aids have the same sensing parameters and most of the babies are multiply handicapped so each is a case study from which preliminary information about use may be obtained. Several researchers are involved and so far none of the work has been published with the exception of that by Bower. It can be said however that of four babies exposed to an aid none have exhibited the spontaneous responses reported by him. During a recent visit to one of the babies reported in the New Scientist (8) it was found that the aid being used by the baby could not be providing even the elementary information it was designed to provide. The experimenter reported failure in the use of the aid. The photograph of the sensory aid shows that this was a modified set of spectacle frames as used for adults and the side arms extended well behind the ears. The sound outlet from each of the side arms was then well behind the ears also and no acoustic coupler to the ears was used. Under these circumstances - as checked with the experimenter - the baby could only hear very loud sounds from specular reflectors; binaural direction cues must have been almost non-existent judging from the work of Keen (35); and the advantage of the wide dynamic range of the ear and its ability to discriminate subtle differences in sound quality was discarded. This gives rise to questions about other reports on babies who used aids which were not properly designed or fitted (50).

     At this time experience suggests a child about to become mobile will appear to benefit most from a binaural sensory aid. Earlier than this the responses so far reported which are supported by the physics of the system do not yet support the hypothesis that a very young brain can readily respond to and assimilate acoustic environmental information. But it may be doing so - when later it may come to use this information more effectively than the older children who are now using aids. The aid in use by some researchers is shown in Fig.21. Special care has been taken to ensure a secure fit so that the sensors are well positioned (they are normally covered by special material as seen in Fig.2) and the sound feed to the ears of a baby remains correct for long periods of time during the day.   Miniature phones as shown have the advantage of creating an "out-thereness" in the sound sensation which otherwise only comes from experience with the distance cue and is self induced. By this is meant that a mental projection of the echo sound comes about as experience is gained in moving with the sensor. This has been verbally reported by adult users. The ears of the baby are clear of sound shadowing caused by the phones which fit forward of the ears so that ambient sound cues can also be learned by a developing baby.

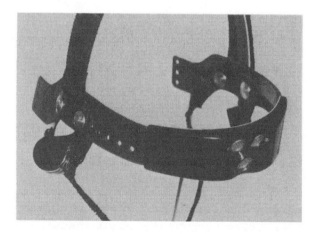

FIG.21.   Babies' Sensory Aid

RELATIONSHIP OF BINAURAL SENSORY AID TO
THE ANIMAL SONAR

It has now been reasonably well established that blind people within the age bracket of approximately 2 to 55 years can learn to interpret, in their own undefined rudimentary way, the rich echo sounds from an ultrasonic sensory aid in which electronic signal processing is minimal (multiplication of reception with radiation). The incoming ultrasonic information is not removed, it is instead transferred to a usable frequency band in the audible range for processing by the human auditory neural system. Appropriate time scaling is achieved in the radiation code. Blind people can learn to perceive and identify certain objects as belonging to a limited set of a given type of object and can negotiate their complex environment with considerable skill. Some locomotion tasks requiring a fine manipulation of auditory flow patterns can be executed with a performance rating which is comparable with that for sighted subjects, and even if only a few blind persons could actually compete with a sighted subject in some of these tasks this is nevertheless performance which two decades ago was not thought likely to be possible. Some experiments with echo-locating animals in the 50's and early 60's as reported at the first NATO Animal Sonar Symposium (Busnel 1966) were really no more sophisticated in relation to animal size and freedom of motion than some of those which have now been carried out with blind humans. Observation of sensory aided blind travellers in a down-town area reveal locomotion skills greatly in excess of those measured in a laboratory but are themselves unmeasurable at present. They cannot even be defined. What then has been learned which can be related to the use of sonar by animals?

The world as it is actually perceived by an echo-locating animal remains unknown to us: up to now it has only been through behaviour patterns that we have guessed at what it must be like. The world as it is perceived by an echo-locating blind person using audible sound bears only a trivial relationship to the real physical world and object identification is negligible (41). Perception is very unreliable. However the world as now perceived through a binaural CTFM sonar does bear strong relationship to the real physical world and anyone with fairly normal hearing can have the same experience as a blind user. It is a unique experience, quite different from that obtained using a high resolution scanning sonar. It may not have been brought about had there not been evidence that animals use echo-location to hunt their prey in complex environments and must therefore be able to resolve multiple object auditory space through two summing apertures.

A great deal has been learned about the ultrasonic reflection characteristics of materials and surface structure of objects

forming our environment in a subjective way which no form of
mathematics or computer modelling could hope to convey; certainly
no simulation system as proposed by the Subcommittee on Sensory
Aids (51) could emulate this. The brain seems to be capable of
categorising a wide range of auditory signals representing objects
and relates these to the particular environment in which it is
operating. The parking meter, for example, can be identified only
if it is located in the environment for which it was designed and
forms an event in a travel sequence. It must be seen as one of
several in a complex pattern of events. A static situation
presented out of context is confusing unless the environmental
structure has been simplified. This is well illustrated in the
evaluation report by Thornton (54) and Armstrong (3). The
influence of ambient cues - sound, smell, etc. - may be considerable
at times.

It is therefore not just the measurable or specifiable
information about the spatial location of objects that the brain
is using, but instead it is the information from the several
sensory modalities which are integrated into a single perception
of the environment as it changes through motion. It is nevertheless
quite evident that the blind are being provided with a simple model
of the animal sonar which can be said to have commonality mainly in
terms of the frequency band of ultrasonic energy radiated into the
environment to be viewed and the reception of echos by two auditory
channels. The coding of the radiated energy has a similarity -
that of a near linear time varying frequency but, whereas the
period of sweep used by a bat may vary from 5 to 0.5 ms, that used
in the sensory aid varies from 250 to 25 ms. The difference arises
from the usable receiver bandwidth of man's sensory system being
only 5 kHz compared with up to 150 kHz enjoyed by bats.

The environmental information which can be extracted by the
two systems - that of the bat and that used by man - may not be
so very different however since the time-bandwidth products have
greater similarity: a 5 ms pulse of sweeping tone radiation
covering an octave frequency band from 100 to 50 kHz and emitted
every 50 ms as may be used by a bat, and a 50 ms pulse of sweeping
tone covering the same octave as may be used in the design of an
aid (see Table 1) have the same time-bandwidth product of 250
when man's limited receiver bandwidth of 5 kHz is used as the
effective handwidth of the system. With this latter system it is
possible to perceive a dead fly tied to a fine thread as it is
swung in and out of the branches of a small bushy fir tree which
may be compared with what the bat perceived in the photograph
shown by Webster (55) of a bat chasing its prey into a similar
shrub tree. The ambiguity function of the CTFM sonar having an
octave bandwidth as discussed by Do (17) helps to explain this
auditory acuity and it is evident that there is no known way in

which this same information can be displayed visually. The
discussion by Do in the next paper is highly relevant.

A similar comparison cannot be made using any other sonic
sensory aids for the Blind: they use a bandwidth of only 1 - 2 kHz
and have a time-bandwidth product of Unity.

Do (17) discusses an underwater sonar using a CTFM radiation
covering an octave bandwidth which has interesting features but
these do not relate to the echo-location by porpoises which employ
quite a different ultrasonic emission. Even so, it seems that
with a CTFM sonar it may be possible to recognise shoals of fish
from other underwater scatterers and it could become possible to
identify different species given adequate experience.

ON AID USE BY THE BLIND

There can be no doubt that blind people need and want a
sensory aid to help them in their everyday activities requiring
perception of their environment, whether this be considered to be
of primary or secondary importance to the Long Cane as a travel
aid matters little. So far, an aid has not been developed which
meets all their declared requirements and it is highly likely
that this goal is beyond man's technical capability. What has
been developed is a range of aids from the simplest - the Long
Cane - to the most sophisticated conceptually - the Binaural
Sensory Aid - with a variety of sonic devices of a simple type,
such as the "Russell Path Sounder", the Nottingham Obstacle
Detector, the "Mowat Sensor" (57) and the Canterbury Single Object
Sensor (33). The Laser Cane also belongs to this latter category.
Of all the promising directions in which sensory aid development
could proceed employing imaging technology and visual implants or
high resolution ultrasonic scanning systems with tactile display,
none actually seem to offer the practical promise inherent in
making most use of the auditory system and its sensitive wide
dynamic range.

This reviewer remains convinced that, as for echo-locating
mammals, humans can make greater use of their auditory system for
sensing their environment. It seems unlikely that the man machine
interface of implanted microelectrodes or a matrix of tactile
stimulators can transfer to the brain the kind of information which
will allow greater perceptual detail for object identification and
motion perception under conditions of mobility. (This should not
be confused with the blind person's need to read or look at single
static objects and the like). But even so, man may not gain
sufficient environmental information from an auditory display to
motivate him to adapt to it regardless of the age of introduction.

The binaural sensor in its further developed form incorporating means to central and peripheral perception, giving greater object recognition in a central narrow field and motion perception in a wide peripheral field, and the means to "focus" on objects through mioelectric control seems to offer the possibility to overcome many of the current limitations of the sensor. This still leaves the need for intellectual stimulus which the lack of vision seriously reduces and spatial auditory perception may not be increased to above the critical useful level which motivates man.

The concept of modelling the bat's sonar so as to make available to man the kind of spatial information enjoyed by the bat fails to take into account man's intellectual needs. The bat is a creature of simple habits, either sleeping or hunting, and an auditory means to catch its prey on the wing in the dusk, remarkable though this may once have seemed to be, does not satisfy man who is already well cared for physically in a highly socialised community. The unknown way in which the bat recognises its environment and can return unerringly to its base poses a more appropriate question. Is the environmental detail much greater than man has available through a sensory aid? Does the pinna of the bat and the flexible use of these apertures give it added information of a kind man would come to use? Unless technological developments can move in these directions we wont learn the answer and this seems to be particularly difficult as shown by Bui (14).

However, the criticism of the Subcommittee on Sensory Aids (51) that the latest techniques in Radar and Sonar systems remain largely untapped in sensory aid development is refuted if it has been shown that the binaural sensor models, as well as it is presently possible technically to model, the bat's sonar and that this model is superior in sensing the environment of a blind man to any other system which can today be proposed. This, at least, would remove a major handicap to acceptance of an acoustic display for a sensory aid but would still leave social attitudes to blindness which determine the economic factors that do perhaps exert the ultimate control on sensory aid use.

REFERENCES

1. Airasian, P.W. "Evaluation of the Binaural Sensory Aid", Research Bulletin No.26, pp.51-71, June 1973.

2. Armstrong, J.D. "Summary Report of the Research Programme on Electronic Mobility Aids", Report, Psychology Dept, University of Nottingham, 1973.

3.   Armstrong, J.D. "An Independent Evaluation of the Kay Binaural
      Sensor". Report to St.Dunstan's Scientific Committee.
      (St.Dunstan's, London, 1972).

4.   Bach-y-Rita, P. "A Tactile Vision Substitution System Based
      on Sensory Plasticity", in (Sterling, T.D., et al. Eds) Visual
      Prosthesis, the Interdisciplinary Dialogue, (Academic Press,
      New York, 1971).

5.   Baird, A.S. "Electronic Aids: Can They Help Children"?
      J.Visual Impairment and Blindness 71, No.3, pp.97-101, March
      1977.

6.   Benham, T.A. "The Bionic Instruments Travel Aid" in
      (Dufton, R. Ed.), International Conference on Sensory Devices
      for the Blind. (St.Dunstan's, London, 1966).

7.   Benjamin, J.M. "The Bionic Instruments C-4 Laser Cane" in
      (Nye, P.W. Ed.), Evaluation of Mobility Aids, (National
      Academy of Engineering, Washington, D.C. 1971).

8.   Bower, T.G.R. "Blind Babies See with their Ears, New Scientist,
      73, pp.255-257, 1977.

9.   Bower, T.G.R. "Babies are More Important than Machines",
      New Scientist, 74, pp.712-714, 1977.

10.   Boys, J.T., Strelow, E.R., Clark, G.R.S. "A Prosthetic Aid
      for a Developing Blind Child", Ultrasonics, 17, pp.37-42, 1979.

11.   Brindley, G.S., Lewin, W.S. "The Sensations Produced by
      Electrical Stimulation of the Visual Cortex". J.Physiol.
      196, pp. 479-493, 1968.

12.   Brabyn, J.A., Strelow, E.R. "Computer-analysed measures of
      characteristics of human locomotion and Mobility", Behaviour
      Research Methods and Instrumentation, Vol.9(5), pp. 456-462,
      1977.

13.   Brabyn, J.A. "Laboratory Studies of Aided Blind Mobility",
      Ph.D. Thesis, University of Canterbury, N.Z.,1978.

14.   Bui, S.T. "A Single Object Sensor as a Mobility Aid for the
      Blind". Ph.D. Thesis, University of Canterbury, N.Z., 1979.

15.   Collins, C.C. "Tactile Vision Synthesis" in (Sterling,T.D.,
      et al.), Visual Prosthesis, (Academic Press, New York, 1971).

16.   Dobelle, W.H. "Current Status of Research on Providing Sight
      to the Blind by Electrical Stimulation of the Brain".
      J.Visual Impairment and Blindness, 71, pp.289-297, 1977.

17. Do, M.A. "Perception of Spatial Information in a Multiple
     Object Auditory Space", Ph.D. Thesis, University of Canterbury,
     N.Z., 1977.

18. Do, M.A., Kay, L. "Resolution in an Artificially Generated
     Multiple Object Auditory Space Using New Auditory Sensations",
     Acustica, 36, pp.9-15, 1976/77.

19. Elliot, E., Elliot, P.H., Roskilly, D. "Sonic Mobility Aid,
     St.Dunstan's Instruction Manual", (St.Dunstan's, London, 1969).

20. Fletcher, H. "Speech and Hearing in Communication"
     (D. Van Nostrand Co.Inc., N.Y. 1953).

21. Gibson, J.J. "Visually Controlled Locomotion and Visual
     Orientation in Animals", Brit.J.Psychology, 49, pp.182-194,
     1958.

22. Guttman, N. (Ed.) "Evaluation of Sensory Aids for the
     Visually Handicapped",(National Academy of Sciences, Wash.
     D.C., 1972).

23. Kay, L. "Active Energy Radiating Systems: Ultrasonic Guidance
     for the Blind" in (Clark, L.L., Ed.) Proceedings of the
     International Congress on Technology and Blindness, (The
     American Foundation for the Blind, N.Y. 1963).

24. Kay, L. "Auditory Perception and its Relation to Ultrasonic
     Blind Guidance Aids". J. Brit. I.R.E., 24, pp.309-317, 1962.

25. Kay, L. "An Ultrasonic Sensing Probe as a Mobility Aid for
     the Blind", Ultrasonics, 2, pp.53-59, 1964.

26. Kay, L. "A Sonar Aid to Enhance Spatial Perception of the
     Blind: Engineering Design and Evaluation". The Radio and
     Electronic Engineer, 44, pp.605-627, 1974.

27. Kay, L., Do. M.A. "An Artificially Generated Multiple Object
     Auditory Space for Use Where Vision is Impaired", Acustica,
     36, pp.1-8, 1976/77.

28. Kay, L. "Evaluation of the Ultrasonic Binaural Sensory Aid
     for the Blind" in (Guttman, N. Ed.), Evaluation of Sensory
     Aids for the Visually Handicapped, pp.85-94, (National
     Academy of Sciences, Wash.D.C. 1972).

29. Kay, L. "Towards Objective Mobility Evaluation: Some Thoughts
     on a Theory". (American Foundation for the Blind, N.Y.1974).

30. Kay, L. "The Sonic Glasses Evaluated", The New Outlook for the Blind, 67, pp.7-11, 1973.

31. Kay, L. "Evaluation of the Ultrasonic Binaural Sensory Aid for the Blind" in (Guttman,N. Ed.),Evaluation of Sensory Aids for the Visually Handicapped, Appendix C, pp.1-33, (National Academy of Sciences, Wash.D.C., 1972).

32. Kay, L., Strelow, E.R. "Blind Babies Need Specially Designed Aids", New Scientist, 74, pp.709-712, 1977.

33. Kay, L., Bui, S.T., Brabyn, J.A., Strelow, E.R. "Single Object Sensor: A Simplified Binaural Mobility Aid", J.Visual Impairment and Blindness, 71, pp.210-213, 1977.

34. Kay, L. "Enhanced Environmental Sensing by Ultrasonic Waves" in (Busnel, R.G., Ed.), Animal Sonar Systems, Tome II, pp.757-781, (Laboratoire De Physiologie Acoustique, Jouy-en-Josas 78, 1966).

35. Keen, K. "Coupling the Output of the "Sonicguide" to the Ear of the User", The New Outlook for the Blind, 70, pp.304-306, 1976.

36. Keith, W.J., "Adaption to Distortion of Auditory Space, M.Sc. Thesis, University of Auckland, 1971.

37. Long Cane; The Cane as a Mobility Aid for the Blind, (National Academy of Sciences, Wash.D.C., 1972).

38. Newcommer, J. "Sonicguide: Its Use with Public School Blind Children", J.Visual Impairment and Blindness, 77, pp.268-271.

39. Nye, P.W. (Ed.), Evaluation of Mobility Aids for the Blind, (National Academy of Engineering, Wash. D.C. 1971).

40. Plomp, R. Steeneken, A.J.M., "Interference Between Two Simple Tones". J.Acoustical Soc. Am. 43, pp.883-884, 1968.

41. Rice, C.E. "The Human Sonar System" in (Busnel, R.G., Ed.) Animal Sonar Systems, Tome II, pp.719-755, (Laboratoire De Physiologie Acoustique, Jouy-en-Josas 78, 1966).

42. Rowell, D. "Auditory Display of Spatial Information", Ph.D. Thesis, University of Canterbury, N.Z., 1970.

43. Russell, L. "Travel Path Sounder" in (R.Dufton, Ed.) International Conference on Sensory Devices for the Blind, pp.293-296 (St.Dunstan's, London, 1966).

44. Saunders, S.D. "The Distortion of an Auditory Display with Noise", M.E. Thesis, University of Canterbury, N.Z., 1974.

45. Scionie, M.W. "Electronic Sensory Aids in a Concept Development Program for Congenitally Blind Young Adults", J. of Visual Impairment and Blindness, 72, pp.88-93, 1978.

46. Sharpe, R. "The Evaluation of the St.Dunstan's Manual of Instruction for the Kay Sonic Aid", Report, Dept of Psychology, University of Nottingham.

47. Stewart, W.G., Hovda, O. "The Intensity Factor in Sound Localization; an Extension of Weber's Law", Psychol. Rev. 25, pp.242-251, 1918.

48. Strelow, E.R., Kay, N., Kay, L. "Binaural Sensory Aid: Case Studies of Its Use by Two Children", J.Visual Impairment and Blindness, 72, pp.1-9, 1978.

49. Strelow, E.R., Boys, J.T. "The Canterbury Child's Aid: A Binaural Spatial Sensor for Research with Blind Children", Journal of Visual Impairment and Blindness [in press 1979].

50. Strelow, E.R., Kay, N., Kay, L. "Binaural Sensory Aid: Response to Comments by Smith and Dailey", J. Visual Impairment and Blindness, 72, pp.320-321, 1978.

51. Subcommittee on Sensory Aids, "Selected Research, Development and Organisational Needs to Aid the Visually Impaired". (National Academy of Engineering, Wash. D.C., 1973).

52. Thornton, W. "The Binaural Sensor as a Mobility Aid", The New Outlook for the Blind", 65, pp.324-326, 1971.

53. Thornton, W. "The Binaural Sensor as a Mobility Aid: An Individual Evaluation of Its Use in Conjunction with the Long Cane", (St.Dunstan's, London, 1971).

54. Thornton, W. "Evaluation of the Binaural Sensor or Sonic Glasses", (St.Dunstan's London, 1972).

55. Webster, F.A. "Interception Performance of Echolocating Bats in the Presence of Interference", in (Busnel, R.G., Ed.), Animal Sonar Systems, Tome I, pp.673-712, (Laboratoire De Physiologie Acoustique, Jouy-en-Josas 78, 1966).

56. Warren, D.H. "Blindness and Early Childhood Development", (American Foundation for the Blind, N.Y., 1977).

57. Wormald Sensory Aids International Ltd., "Mowat Sensor" a commercial product. Christchurch, New Zealand.

DISCRIMINATION OF COMPLEX INFORMATION IN AN ARTIFICIALLY

GENERATED AUDITORY SPACE USING NEW AUDITORY SENSATIONS

Manh Anh Do

Radio Engineering Limited

P.O. Box 764, Dunedin, New Zealand

INTRODUCTION

Wide beam broad band CTFM (continuous transmission frequency modulated) sonar with auditory binaural display was first proposed to be used as a sensory device for the blind person by Kay [16] in 1959. In this system the combined distance and direction of an object is coded in the form of rising pitch with increasing distance (frequency proportional to distance) and the binaural differences which could include that of time, amplitude, and frequency. Subsequent research to optimise this display resulted in a reduction of the interaural differences to that of amplitude only, and this difference measured in dB was designed to be proportional to the azimuthal angle as far as was physically possible [17, 28]. Thus the CTFM sonar with the binaural display (or briefly the binaural sonar) basically performs a one-to-one transform of the two dimensional real space-distance (range) and direction (azimuthal angle) of an object into a two dimensional auditory space described by frequency and interaural amplitude difference (IAD) of.a tone. This auditory space was readily produced by means of a sensory aid for the blind [24, 18], and a sonar to locate fish [33, 34]. The evaluations of these two systems were reported in [2, 19, 34, 35, 5]. The results indicated clearly the ability of human operators to use the new display and to perform their tasks successfully with relative ease, even though the input to the ears was rich in information.

The results obtained from field evaluation were however impossible to explain by the experimental results found by Rowell [28] and Anke [1]. The latter was from laboratory experiments

using system simulation designed to determine the ability of
subjects to resolve objects in space by auditory means. The
results of Rowell and Anke suggested a very poor auditory resolution.
The disagreement between these two results led the author and Kay
to the investigation of the role of Doppler effect in the binaural
sonar [20, 4, 5]. Rowell and Anke's experiments used only static
situations to facilitate controlled measurement. The temporal
effects and the complexity of the sounds produced by the binaural
sonar under dynamic conditions, which could possibly enhance the
resolution capability of subjects were neglected in their
experiments.

This paper reviews the experimental results reported in [5]
on the capability of subjects to discriminate complex information.
New auditory sensations were found to be perceived by subjects in
laboratory experiments using system simulation close to realistic
situations. This resulted in the high performances of the users of
the binaural sonars as described in [34, 2, 19]. The experimental
results were related to the theoretical research on the hearing
system and the sonar system to evaluate the performance of the
man-machine system as a whole.

THE AUDITORY FREQUENCY RESOLUTION OF TWO PURE TONES SIMULTANEOUSLY SOUNDED

Since the audible information displayed by the sonar, when a
multiple of objects is presented in the field of view, is a large
number of tones, each with its own characteristics, the first
approach is to determine the auditory resolution of two tones.
Several different psychological problems are however immediately
arisen under such simple conditions of simulation. Indeed, even
if the two tones are monaurally presented, depending on the
combinations of four variables (two amplitudes and two frequencies)
there will be different effects, namely, beats, combination tones
(CT's), masking, fusion, discrimination, etc., on subject's
perception.

In 1924, Wegel and Lane [41] designed a very elegant
experiment to measure the shift of the threshold of hearing of a
pure tone when masked by another. However, when using the 80 dB
1200 Hz tone as the masker and increasing the level of the masked
tone beyond its threshold of hearing up to the same level of the
masker, over the range from 400 to 4000 Hz, they could determine
the limits in intensity and frequency of the masked tone where
either beats, or combination tones, or one tone only, or two tones
only, etc., would occur (figure 1). Their results showed that a
tone can mask tones of higher frequency more effectively than those
of lower frequency than itself. Wever [42, 43] described the beat

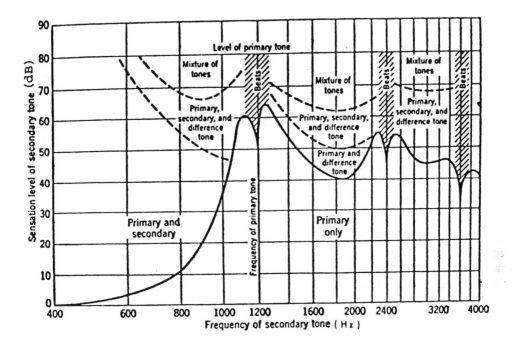

Figure 1.   Criteria of various sensations caused by the
             simultaneous sounding of two tones. The primary is the
             80 db, 1200 Hz tone. The frequency of the secondary
             varies from 400 to 4000 Hz and its level from 0 to 80
             db (Wegel and Lane 1924).

phenomenon by three clearly distinguishable stages, appearing
successively as the frequency difference of two tones is increased
from zero-(i). Noticeable oscillation: the loudness rises and
falls slowly; (ii). Intermittance or pulsation of tone: pulses of
tone separated by silence; (iii) Roughness without intermittance:
there appears a sort of whirr, and the sound is characteristically
rough. Observing two tones of equal sensation levels of 20 dB, one
was at 1024 Hz while the other was varied from that frequency upward,
Wever found the above three stages of the beat phenomenon at the
frequency differences of: (i) 0 to 6 Hz, (ii) 6 to 166 Hz, and
(iii) 166 to 356 Hz.   All these results [41-43] show that a
frequency difference of several hundreds Hz is required so that
subject can hear two tones, say, of 1000 Hz and 1000 Hz + Δf,
without being interfered by the combination tones. These
experiments were however not extended to different frequencies.

The purpose of this study is to seek for a definition of the auditory frequency resolution which is related to the spatial range resolution in a sense that: for two single objects presented in the field of view of the sonar system, the listener will hear as if only two tones are presented and the perceived tones must be the coding tones displayed by the sonar. Plomp et al [25-27] defined four criteria: (i) Frequency difference between partials of a complex of tones required to hear them separately, (ii) frequency difference between two tones required to distinguish two pitches, (iii) frequency difference between two tones for maximum harshness, and (iv) frequency difference for just absence of beats. Investigating the results shown in figure 2 and the experimental procedure described in [25-27] it is believed that the fourth criterion is equivalent to the end of the second stage of beats described by Wever [42, 43] while the third is the maximum perceptible intermittence of this stage. A forced choice procedure was used in the first and second criteria. Subject was asked to compare a complex stimulus with two different single tones successively, and to choose the single tone which belonged to the complex stimulus. This procedure does not testify to either the absence or the existence of the combination tones. So neither of these criteria are able to define the resolution of the auditory space generated by the binaural sonar. Using two tones having different IAD's, Rowell [28] found that subject requires a fractional frequency difference, $\Delta f/f$, of 40% to tell which tone, low or high, is in the left or in the right. This result seems not too dependent on the IAD's of the tones.

In [4,5] the author defined the frequency resolution as the least frequency difference between two tones so that they are perceived as being "distinctly separate" where the term "distinctly separate" implies the absence of the interference between two tones so that the stimuli are the only perceived tones. This ensures a one to one transformation from the real space to the auditory space generated by the binaural sonar. Two tests were used to verify the subject's judgments at the criterion of resolution.

(i) Two and only two tones are perceived by subject, and

(ii) The perceived tones are the stimuli.

In test (i) subject is presented with a stimulus which randomly includes either two or three components, and has to answer the question "is this only two tones"? Immediately after a correct "yes" answer is given, test (ii) is continued. Subject is presented with a single tone, and he has to answer either of the two questions, "is this the low tone"? or "is this the high tone"?, while the single tone may be either the low tone, the high

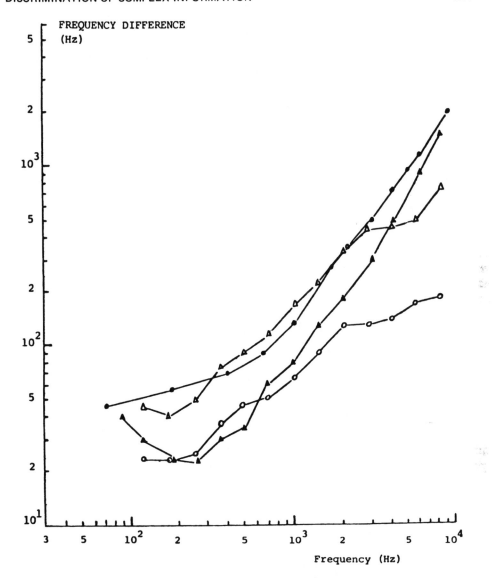

Figure 2.   Four Criteria of Sensations defined by Plomp et al.

● Frequency difference between the partials of complex
    tones required to hear them separately.

▲ Frequency difference to distinguish two tones
    simultaneously presented.

o Frequency difference for maximal roughness of two tones.

△ Frequency difference for just absence of beats given
    by two tones.

tone, the difference tone, or the cubic difference tone. Experiments with four subjects shows an average frequency resolution of $\Delta f/f = 40\%$. Below this limen, the perception of the fundamentals are interfered by the combination tones considerably.

ROLE OF DOPPLER EFFECT IN THE BINAURAL DISCRIMINATION OF AUDIBLE
FREQUENCY PATTERNS PRODUCED BY WIDE BAND CTFM SONAR

The transmitted signal of a CTFM sonar system is cyclic and of the form:

$$S_T(t) = A \exp [2\pi j (f_2 t_s - \frac{m}{2} t_s^2)] \tag{1}$$

Where f_2 = upper limit frequency

m = sweep slope

t = $t_n + t_s$ at any instant, t_n is the beginning time of the n-th cycle of the modulating function, and t_s varies from 0 to T_s, the repetition period.

A = constant amplitude of the transmitted signal.

Let $r(t)$ be the instantaneous range of an object. The received signal, $S_R(t)$, is then a Doppler shifted replica of the transmitted signal, $S_T(t-\tau(t))$, $\tau(t)$ seconds ago, which was reflected from the object at distance $r(t_r)$, where $t_r = t - \frac{\tau(t)}{2}$. Thus

$$\frac{\tau(t)}{2} = \frac{1}{c} r(t_r) = \frac{1}{c} r(t - \frac{\tau(t)}{2}) \tag{2}$$

where c is the propagation velocity of sound.

Substituting (2) in $S_T(t-\tau(t))$, the range coding frequency of an object is the frequency difference of the transmitted and received signals.

$$f_a(t) = m\tau(t)(1 - \dot\tau(t)) - \dot\tau(t)(f_2 - mt_s) \tag{3}$$

In perfectly stationary conditions, the range of an object is constant, thus the delay time is:

$$\tau = \frac{2r}{c} = \text{constant, and } \dot\tau = 0 \tag{4}$$

Hence $f_a = \dfrac{2m}{c} r = Kr$ (5)

The constant $K = \dfrac{2m}{c}$ is the range code of the sonar. With a
suitable choice of the sweep slope m, the range coding frequency,
f_a, can be made audible. Typically, the sensory aid for the blind
has a range code of 941 Hz/metre, and the fishing sonar has four
different range codes of 53.33, 26.67, 13.33, and 6.67 Hz/m for
four different maximum operating ranges of 94, 188, 375 and 750m.

Under dynamic conditions, the range of an object is no longer
simply proportional to the coding frequency of the displayed sound
as described by relation (5). To solve equations (2) and (3), the
simplest way is to approximate $\tau(t)$ by:

$$\tau(t) = \dfrac{2r(t)}{c}$$ (6)

this will meet an error of

$$\dfrac{\Delta\tau(t)}{\tau(t)} = \dfrac{1}{c} \dfrac{r(t) - r(t - \tau(t)/2)}{\tau(t)/2} = \dfrac{\dot{r}(t)}{c}$$ (7)

However, while the fishing sonar is being operated, the
operator steering the boat does not travel faster than 3m/s (six
knots), hence $\dot{r}(t)/c$ is smaller than 0.2%, where c = 1500 m/s is
the sound velocity in water. A blind user of the sensory aid
walks at a speed, at most, 5 km/hour. The average sound velocity
in air is 340 m/s. Thus $\dot{r}(t)/c$ is smaller than 0.4%. The
approximation in equation (6) is reasonable. Replacing (6) in
(3) we have:

$$f_a(t) \doteq \dfrac{2m}{c} r(t) - \dfrac{2\dot{r}(t)}{c} f_T(t)$$ (8)

where $f_T(t) = f_2 - mt_s$ is the instantaneous transmitted frequency.

Hence the relative motion between object and sonar system
basically changes the frequency of the sonic wave by a factor
$(1 - 2\dot{r}(t)/c)$. The frequency change $\dfrac{2\dot{r}(t)}{c} f_T(t)$ is called the
Doppler shift. Equation (8) shows that the Doppler shift is
preserved in the range coding frequency of the CTFM sonar. An
example of the audible frequency pattern produced by a fishing
sonar due to constant radial velocity $\dot{r}(t) = -v$ is shown in
figure 3.

$$f_a(t) = \dfrac{2m}{c} (r(0) - vt) + \dfrac{2v}{c} (f_2 - mt_s)$$ (9)

This sawtooth cyclic variation pattern is due both to the
Doppler shift of frequency in the medium, and the wide variation
in transmitted frequency.

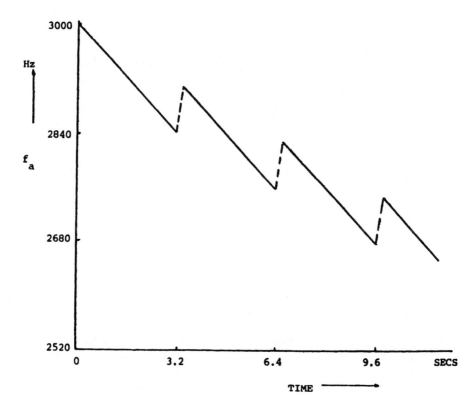

Figure 3. Variation of Range Coding Frequency with Respect to
          Time.

$$f_{a1}(t) = 3000 - 25t - 25\ t_s, \text{ where } 0 < t_s < 3.2.$$

The broken lines correspond to ambiguous signals
blanked out.

Binaural Sound Pattern

        The discussion has thus far been about the range indication.
Under realistic conditions, the relative direction of the object
is also changing during locomotion of the blind aid user.   Hence
the audible frequency pattern will flow left or right in
accordance with the changing spatial positions of the object
relative to the user.   The realisation of the direction code is
shown by Kay in the same chapter.

        Two typical situations encountered in realistic conditions --
(i) object passed in one side as subject moves along a straight
course, and (ii) subject rotates his head to obtain more
information - were analysed by the writer in [5].   Figure 4 shows

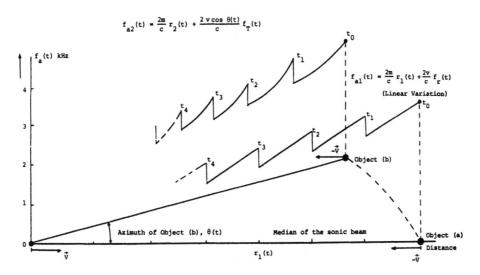

Figure 4.   Comparison between the sound patterns of a centre
object and of a side object.

two different frequency patterns perceived by a user of the sensory
aid while moving along a straight course. One is produced by a
central object, and the other due to an object on the left side.
The possibility for a sensory aid user to discriminate two objects
different in direction but close in range, by rotating his head, is
illustrated in figure 5. The Doppler effect mainly creates a
negative frequency pulse when the user turns away from the object,
and a positive frequency pulse when he turns towards it.

Hence under real life conditions, for example, mobility by the
blind in a busy pedestrian area where people are moving with
varying acceleration from different directions, and the relative
velocity between user and fixed objects will also vary, the audio
signals from the sensory system will change in an indescribably
complex way, each signal having its own unique character.
Experience with blind people using the sensory aid suggests that
great resolution of these time varying tones is being enjoyed.
What they perceived are the patterns of sound, which flow towards
the left or the right in a unique way related to the changing
positions of the reflecting objects.

In the case of the fishing sonar, similar sound patterns
could be produced due to motion of either the boat or the targets.
Two kinds of motion of shoal target could be considered:
(i) translational motion of the entire shoal and, (ii) oscillatory
motion of individual fish within the shoal. The translation
velocity of the shoal is frequently smaller than 75 cm/s [13, 23],

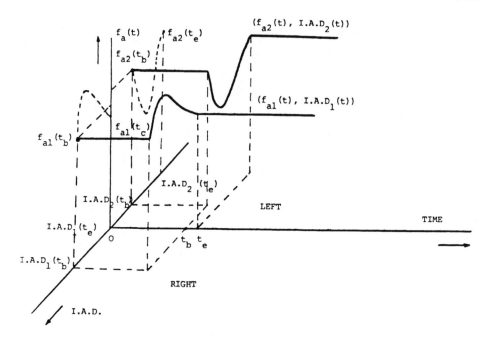

Figure 5.   The effect of head rotation on the binaural sound
             patterns of two objects initially close in range but
             different in azimuthal angle (head rotates to the
             right in this figure).

the equivalent frequency shift is less than 60 Hz. However the
sound pattern created by the shoal when it crosses the sonic beam
from left to right or vice-versa is not unlike that obtained in
the sensory aid when subject rotates his head. Bainbridge [3]
showed that average fish (length 75 cm) can reach a speed of 6 m/s
though only for a short period. Hence the oscillatory motion of
fish in the shoal can create the frequency shifts up to 480 Hz in
the echoes if 60 kHz signals are transmitted. Using 70 kHz
transmitted signal Hester [15] observed the frequency shifts of
about 200 Hz from individual fish. In addition to this effect,
amplitude modulations in the sound pattern (due to the changes in
scattering cross section of fish while they are swimming) are also
possibly heard. Fish of different sizes or different kinds have
different types of motion. Hence the sound patterns they produce
must be different. An experienced user of the binaural sonar will
be able to learn to recognise different species from the difference
in character of the sound patterns these species produce.

A DEFINITION OF FREQUENCY RESOLUTION USING NEW AUDITORY SENSATIONS

Consider two adjacent single objects, 1 and 2, described by their instantaneous ranges $r_i(t)$ and azimuthal angles $\theta_i(t)$ where $i = 1,2$ in the field of view of a binaural sonar. Correspondingly, in the auditory space the two time varying tones will be described by:

$$f_{ai}(t) = \frac{2m}{c} r_i(t) - \frac{2f_T(t)}{c} \dot{r}_i(t) \qquad (12)$$

$$IAD_i(t) = 10 \log \frac{I_{Ri}(t)}{I_{Li}(t)} = K_1 \theta_i(t) \qquad (13)$$

where $i = 1,2, I_{Ri}$ and I_{Li} are respectively the sound intensities in the right and left ears of the i-th tone. Constant K_1 is determined by the equation (11). Typically it is 0.4 to 0.5 dB/ degree in the sensory aid, and about 1.5 dB/degree in the fishing sonar.

Thus the capability of resolution must be a function of four dynamic variables $[f_{a1}(t), IAD_1(t), f_{a2}(t), IAD_2(t)]$. Clearly any function of four dynamic variables is too complicated to be studied and described by subjects in a psychophysical experiment. In all experiments described in (12) and (13) $f_{a2}, IAD_2,$ and IAD_1 were held constant, only f_{a1} was varied as in equation (12). The procedure was based entirely upon the "psychological phenomenon" described below.

The Psychological Meaning of Auditory Resolution Under Conditions of Change

It was known in Section 2 that under static conditions the assignment of different values of IADs to the two pure tones did not affect the frequency resolution, but it defined the psychological phenomenon of frequency resolution more clearly, since subjects were required to indicate the left or right directions of the tones. This procedure was adapted to the dynamic experiment as follows:

Initially at t=0, and $f_{a1}(t) \doteq f_{a2}$ the subject hears slow beats, and feels only one complex image in the auditory space. Its IAD is the combination of IAD_1 and IAD_2. As t increases $f_{a1}(t)$ decreases according to equation (12). The interference between two

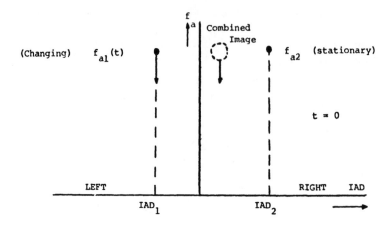

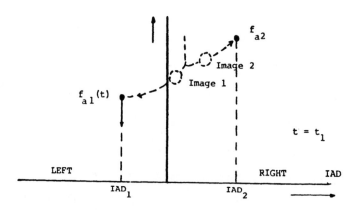

Figure 6.   The Psychological Phenomenon of Resolution in a
                Changing Situation

tones changes in a unique way until a certain time $t=t_1$, the
subject feels the image of $f_{a1}(t)$ flow "left" or "right" in auditory
space according to the assigned values of IAD_1 and IAD_2 . The
attention appears to be captured by the "moving" tone which may be
thought of as forming a <u>flow pattern</u> in space. At this instant the
subject is required to respond by pressing a switch to store the
value of $f_{a1}(t)$. $f_{a1}(t)$ is allowed to continue when the attention
can be transferred to f_{a2} which then appears in its appropriate
auditory spatial position. This is always secondary to the
initially perceived flow of $f_{a1}(t)$, but occurs almost

simultaneously. The psychological phenomenon is described in figure 6. The judgment of subjects can be verified by questioning them on the relative movement of the images. The frequency resolution can be defined here as the frequency difference between f_{a2} and $f_{a1}(t)$ at which the two tones separate and "slide" in auditory space to their respective positions.

Measurement of Frequency Resolution

The simulation related to a fishing sonar was chosen in the experiments because the rates of change encountered are such as to allow a subject to respond without introducing serious error due to the delay in response. The situation of constant radial velocity was simulated.

Typically $v = 1.5$ m/s, $c = 1500$ m/s, $T_s = 3.2$ secs, $m = 12.5$ kHz/s, thus

$$f_{a1}(t) = f_{a1}(0) - 25(t + t_s) \tag{14}$$

where $0 \leqslant t_s \leqslant 3.2$s.

The variation of $f_{a1}(t)$ with respect to time is shown in figure 3. Values of $f_{a1}(0) = f_{a2}$ were chosen to be 1000, 2000, and 3000 Hz, while the values of IAD were 0, +4, +8, + 12 dB for tone 1, and −10, −6, −2, +2, +6, +10 dB for tone 2. Figure 7 shows the results obtained from three subjects S_1, S_2, and S_3. The variation of $\Delta f_a = f_{a2} - f_{a1}(t_1)$ with respect to different values of IAD_1 and IAD_2 was found to be smaller than the possible error due to the delay of subject's reaction. Hence the dependence of frequency resolution on IAD could be considered as negligible.

From the above results it is obvious that the capability of frequency resolution, under dynamic conditions is significantly improved when compared with that under stationary conditions. Equation (14) shows that the Doppler effect results in an increase of the rate of change of the range coding frequency by a factor of two. This also takes part in the improvement of the frequency resolution. Indeed by removing the cyclic variation in the tone $f_{a1}(t)$ so that

$$f_{a1}(t) = f_{a1}(0) - 25 t \tag{15}$$

and repeating the above experiment with subject S_2, his capability of resolution was found to be worsened (see table 1).

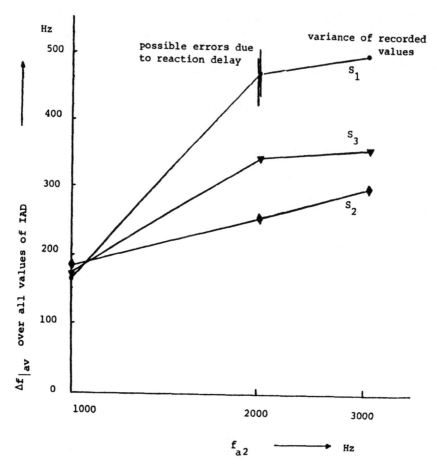

Figure 7.   Variation of the frequency resolution with respect to
             frequency averaged over all values of IAD (IAD variable
             has negligible effect).

          Since the cyclic variation involving doubling the rate of
change of the audio frequency within each sweep, has a significant
effect on the resolution capability, a direct increase in the
object velocity, resulting in a higher rate of change of frequency
has to have certain effects on the resolution performance of the
subject. By increasing the object velocity to 3 m/s, so that

$$f_{a1}(t) = f_{a1}(0) - 50 \ (t + t_s) \eqno(16)$$

and repeating the experiment with subject S_2, his resolution
capability was found to be improved quantitatively as well as
qualitatively (see table 1). The sensation of image slide in
auditory space is more pronounced and the separation of images is

very clear by all who experience the phenomenon. Subject S_2 was unable to respond quickly to change greatly the measured resolution capability obtained in the experiment using equation (14) for the simulation. Hence in the conditions of high rates of change of audio frequencies as in the sensory aid for the blind, signals must be resolved cognitively long before reaction can be recorded.

TABLE 1.   Comparison of Frequency Resolution of Subject S_2 in four different conditions.

f_{a2} Hz	$\Delta f(0)$ Hz	$\dfrac{\Delta f(0)}{f}$	$\Delta f(1)$ Hz	$\dfrac{\Delta f(1)}{f}$	$\Delta f(2)$ Hz	$\dfrac{\Delta f(2)}{f}$	$\Delta f(3)$ Hz	$\dfrac{\Delta f(3)}{f}$
1000	400	0.50	180	0.20	220	0.25	130	0.14
2000	700	0.42	250	0.13	440	0.25	230	0.12
3000	1100	0.45	300	0.11	630	0.23	240	0.08

$f(0)$: stationary tones

$f(1)$: saw-tooth variation of $f_{a1}(t)$

$f(2)$: Linear variation of $f_{a1}(t)$

$f(3)$: saw-tooth variation of $f_{a1}(t)$ at double rate

AUDITORY DISCRIMINATION OF TWO MULTIPLE COMPONENT TONES

The investigation on the multiple target resolution of the binaural sonar so far has been carried out with single objects only. Under realistic conditions a target always has a certain size, and is characterised by its shape. Hence the corresponding audible signals displayed by the binaural sonar at its two channels must be the band limited tones whose spectra are combined in a specific way in accordance with the shape of the target. To describe a target consisting of, say, n "scattering centres" (locations scattering most of the power back to the receiver) n pairs of variables $(R_i(t), \theta_i(t))$ are required, where i=1, n, and $R_i(t)$ and $\theta_i(t)$ are respectively the instantaneous range and azimuth of the i-th scattering centre. Correspondingly in the auditory space, the multiple component coding tone, or simply the complex tone, must be described n pairs of variables $(f_{ai}(t), IAD_i(t))$, where $f_{ai}(t)$ and $IAD_i(t)$ are determined by equations (12) and (13), and i=1, n. However to simplify the mathematical presentation of the problem, it will be assumed that

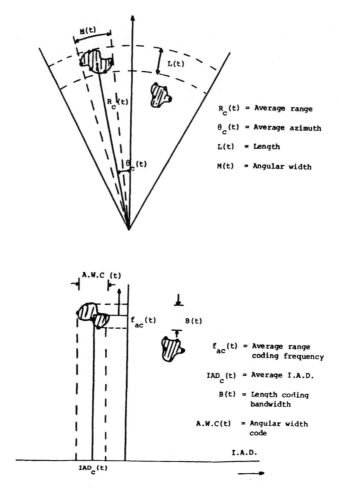

Figure 8.   Relationship between the variables in real space and
             in auditory space.

within a short duration of time, when a certain task such as
discrimination of two adjacent targets is carried out, the sizes
and shapes of the targets are unchanged. Consequently each target
can be simply described by two variables, its average range $R_c(t)$
and azimuth angle $\theta_c(t)$, and two parameters, the extent in range
(or length) L and the angular width M of the target. Figure 8
illustrates the one to one relationship between the above variables
and parameters in real space with the variables $f_{ac}(t)$ (average
range coding frequency) and $IAD_c(t)$ (average IAD), and parameters
B (length coding band-width) and A.W.C. (angular width code) of
the multiple component tone in the auditory space.

     A ten channel voltage controlled oscillator was especially
built to synthesize two complex tones, each of five equally spaced

components. The experimental procedure was based upon the psychological phenomenon similar to that found in the experiment with single tones described in Section 4. The most surprising result is that the discrimination of two multiple component tones under certain conditions is easier than the discrimination of two single tones.

Initially the length coding bandwidths of the two complex tones were set at 100 Hz, and the angular width codes were 2 dB. The values of $g_{ac} = f_{ac}(0)$ were chosen to be 1000, 2000, and 3000 Hz while IAD_c were 0,4,8 dB, and JAD_c were −10, −6, −2, +2, +6 and +10 dB. For each combination of g_{ac}, IAD_c, and JAD_c the recording frequency discrimination in four times showed little deviation, frequently less than 20 Hz. An average $\dot{f}_{ac} = 130$ Hz at $g_{ac} = 1000$ Hz at $g_{ac} = 2000$ Hz, and 240 Hz at $g_{ac} = 3000$ Hz. Figure 9 shows very close results given by three subjects. Besides the better discrimination performance obtained in this case of complex tone in comparison to the performance with single tones, it was noted that the frequency discrimination was slightly worsened when IAD_c and JAD_c are close together. Although this increase of Δf_{ac} was not very large, it was consistently found in the performance of all three subjects at different frequencies, and IAD's.

In order to study the effect of the angular width code on the frequency discrimination, the above experiment was repeated with subject S_2 using complex tones having angular width code increased to 4 dB. The bandwidths of the tones were still kept at 100 Hz. Δf_{ac} is increased by up to 20% in this case. Also there is always a small increase in Δf_{ac} as the difference between IAD_c and JAD_c decreases.

The length coding bandwidths of the complex tones were then increased to 200 Hz, while their angular width codes were restored to 2 dB. Repeating the experiment with subject S_2, it was found that in comparison to the results obtained in the experiment using 100 Hz bandwidth tones, Δf_{ac} increases 79% at $g_{ac} = 1000$ Hz, 65% at $g_{ac} = 2000$ Hz, and 69% at $g_{ac} = 3000$ Hz. At $g_{ac} = 1000$ Hz, subject reported the difficulty in separating two tones.

DISCUSSION

From Basilar Membrane Filtering to Frequency Analysis of Collicular Neurones

The "cochlea" in the inner ear is the primary frequency selective part of the hearing organ. Von Békésy [6] discovered

M. A. DO

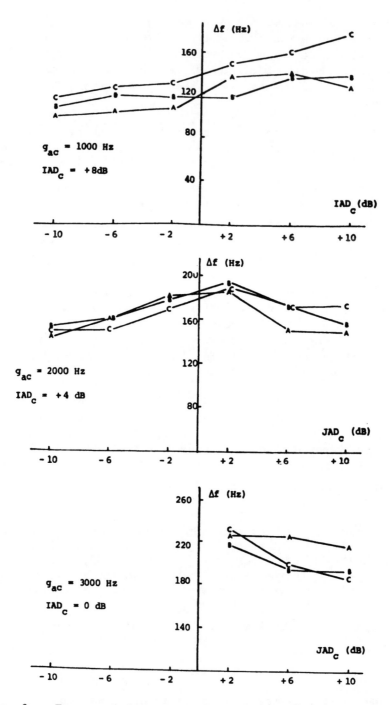

Figure 9.   Frequency discrimination of two complex tones as a
             function of I.A.D. performed by three subjects A, B,
             and C.

the "travelling waves" on the "basilar membrane" inside the cochlea
and recognised it as a "non-uniform" transmission line where lower
frequencies can travel further than higher frequencies from the
"stapes" before being attenuated. This observation led to the
"place theories" of pitch perception, according to which the
position of maximum vibration of the basilar membrane determines
the pitch of the tone. The frequency resolution of the basilar
membrane as observed by Von Békésy was very low. Two tones
simultaneously sounded must be separated by a fractional difference
of $\Delta f/f \doteq 30\%$ before usual fatigue effects appear. He considered
this as the limen for two tones to be heard distinctly. Another
approach to study the frequency selectivity of the ear was the
introduction of the concept of the "critical band" [6,8,12,14,29,
30,39,45,46] as : (i) either the limited band of noise effectively
masking a tone, (ii) or the frequency band within which the energy
of a multiple component tone is summated, etc. Widely diverse
results were obtained from these authors. $\Delta f/f$ is from 10% to 25%.
Several experimenters have had the tendency to compare these
critical bandwidths with the frequency resolution capability [25-27],
while it has been shown in section two that two single tones
simultaneously sounded must differ considerably in frequency,
$\Delta f/f \doteq 40\%$, to be perceived distinctly. However, if we consider the
fact that the ear can easily detect a frequency change of $\Delta f/f$ less
than 0.5% with respect to time, either abruptly or in a certain
continual mode [7,35,44], can track a fast time varying tone (as
experienced by the sensory aid users), and can resolve time varying
tones better than stationary tones, then the assumption of a
mechanical filtering process in the cochlea as the only frequency
analysis is apparently inadequate.

To understand the above unconformable results, let us postulate
that: The inner ear can perform a real time frequency analysis of
the stimuli but sustains a high degree of distortion. The assumption
of a real time frequency analyser mechanism explains the capability
of the ear in tracking a fast time varying tone as well as detecting
a very small frequency change with respect to time. However when
the stimulus consists of two or more components, the perception of
the fundamentals is considerably interfered by the distortion
products, unless the fundamentals are well separated from each
other.

Considering now three most prominent subjective components
heard by a subject when he is presented with two adjacent tones,
f_{a1} and f_{a2}, simultaneously in three different cases (Table 2):

It is obvious that while all three subjective components are
stationary in case 1, in case 3 only the high tone is stationary,
the cubic difference tone descends at a rate of change in frequency
equal to twice that of the low fundamental tone. This can be

TABLE 2.   Three most prominent subjective components produced
by two tones f_{a1} and f_{a2} simultaneously presented.

Subjective component	case 1 stationary tones	case 2 $f_{a1}(t)$, cyclic variation	case 3 $f_{a1}(t)$, linear variation
High tone	f_{a2}	$f_{a2}=f_{a1}(0)$	$f_{a2}=f_{a1}(0)$
Low tone	f_{a1}	$f_{a1}(0)-qt-qt_s$	$f_{a1}(0) - qt$
Cubic difference tone	$2f_{a1}-f_{a2}$	$f_{a1}(0)-2qt-2qt_s$	$f_{a1}(0) - 2qt$

where $q = \dfrac{2v}{c}$ m, t_s varies from 0 to T_s

generalised that under dynamic conditions, if the stimuli have
different rates of change in frequency, all the subjective
components will vary at the rates of change in frequency which are
entirely different from each other. Hence it is possible that at
higher levels of perception, following the filtering process of the
basilar membrane, the collicular neurones respond to these
subjective components differently, so the interference of the
combination tones in the perception of the fundamentals is less
significant than that in the static conditions when both the
fundamentals and the combination tones are of the same
characteristic: "STATIONARY". Accepting this hypothesis the
explanation for the higher resolution obtained in case 2 becomes
obvious. The Doppler effect in the CTFM sonar increases the rates
of change of the range coding frequencies, increasing also the
differences between the rates of change in frequency of the
subjective components. The same argument applies to the case of
direct increasing in the object velocity.

Little knowledge about this performance of human beings was
found in literature. Suga (1965, 1968) [36-38] studied the
responses of collicular neurones of bats to frequency modulated
(FM) and complex sounds, classified them in five main types:-
symmetrical, asymmetrical, FM sensitive, FM insensitive, and
upper threshold units - depending on the way they respond to their
best frequencies, the inhibitory areas which limit the ranges of
frequency modulation on one or both sides of the best frequencies,
the selective responses of the units to certain types of modulation

determined by the range, rate, direction and functional form of the frequency sweep, and the thresholds of intensity to activate the neurones.   These principles were believed to apply to all kinds of mammals and human beings but of course with different time scales. The characters of selective responding of the collicular neurones explain the capability of the auditory system in analysing the structure of a complex sound. Hence under realistic conditions, the time varying tones produced by a binaural sonar are furnished with additional characters such as rates of change in frequency, functional forms, etc. which enable the auditory system to discriminate the fundamentals from the distortion products more easily.

The Improvement and the Worsening of Frequency
Discrimination to Multiple Component Tones

It was known from place theory that if the stimulus consists of several frequency components which are not separately resolved, then an averaging mechanism does take place [21,22], a pitch of frequency approximately equal to that of the centre component will be heard. An explanation for the improvement or the worsening of the frequency discrimination of the multiple component tones in comparison to that of single tones can be based upon that averaging mechanism as follows.

Throughout the discussion in the previous section it was seen that the CT's, especially the cubic difference tone $2f_1 - f_2$, generated due to the non-linearity of the inner ear, interfere significantly in the discrimination of the fundamentals. Figure 10a shows the typical level of the $2f_1 - f_2$ tone when f_1 and f_2 are equal to 70 dB [9-11]. However if two complex tones of the same levels (70 dB) as the single tones, and having the same bandwidths B's, are now considered, the cubic difference tones generated by the combinations of various pairs of components will be spread out over a large frequency band as demonstrated in Figure 10b. This also applies to higher order CT's.   Therefore if the bandwidths of the complex tones are narrow enough so that the averaging process can take place, two stimuli must raise two pitches of frequencies equal to the centre frequencies of the complex tones and of intensities equal to that stimulated, while all the CT's are spread out and create a low level wide band background. The interference of this background may not be as considerable as the interference of the CT's generated by a two pure tone stimulus.

Accepting the above explanation for the improvement of frequency discrimination of narrow band (100 Hz) tones leads to the only possible explanation for the worsening of frequency

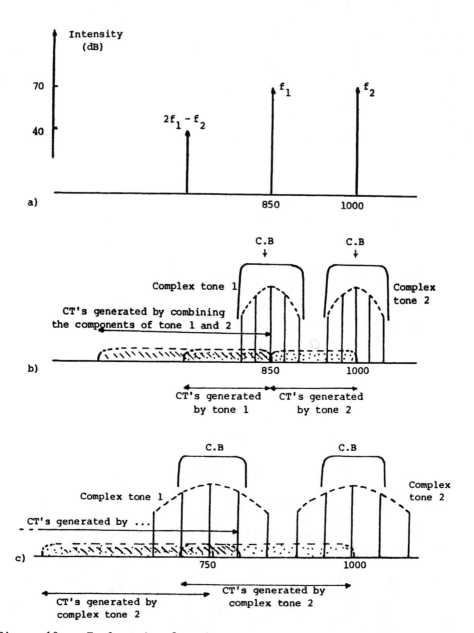

Figure 10.   Explanation for the improvement of frequency
              discrimination of narrow band tones and the worsening
              of frequency discrimination of wide band tones.

discrimination of wide band (200 Hz) tones that the averaging
mechanism fails to summate the components of the complex tones
effectively in the latter case, the stimulus tones are diffused and
the $\overline{CT}$'s generated by the high complex tone interfere seriously
on the low tone (Figure 10c). Zwicker et al [45,46] and Sharf
[29,30] measured the critical band within which the power of a
complex tone is summated and found that this bandwidth could vary
from 80 Hz at f=100 Hz to 2500 Hz at f=10,000 Hz. According to
their results the critical handwidth is equal to 160 Hz at
f=1000 Hz, 300 Hz at f=2000 Hz, and about 500 Hz at f=3000 Hz.
Subject S_2 reported the diffiuclty in separating two tones at
g_{ac} = 1000 Hz because the bandwidths of the two complex tones were
larger than the critical bandwidths. At g_{ac} equal to 2000 Hz and
3000 Hz, the summation of the power of each complex tone may not
fail, but the CT's generated by the high tone still interfere on the
low tone considerably. The discrimination is worsened but the tones
are still easily perceived and separated.

CONCLUSION

Throughout this paper, results obtained from several
psychophysical experiments using system simulation techniques are
described. An investigation of the frequency discrimination
performance in a multiple object auditory space was approached from
the very simple situation of two single stationary tones to a much
more complicated situation of two multiple component tones whose
frequency difference varied with respect to time. The experimental
results have explained why a human operator was able to use the new
auditory display in realistic conditions with relative ease, as
observed in field evaluation, even though the inputs to the ears
were very rich in information.

This paper is a review of four chapters of the complete work
[5] whose purpose was to evaluate an auditory display for use with
the broad band CTFM sonar. The actual resolutions in range and in
velocity of the CTFM sonar itself are much higher than the
resolutions of the display. In [5] the ambiguity function of the
broad band CTFM signal was determined and applied in the example
of the fishing sonar described in section 4. In this approach, the
resolutions in range and velocity are defined by the limits where
the cross-correlation between the echoes returned from two adjacent
targets is equal to a quarter of the maximum value when the targets
are completely coincident. The maximum values of the resolutions
in range and velocity were found to be $\Delta r \doteq 0.375$ ft, and $\Delta v \doteq 4 \times 10^{-2}$
ft/s. These values correspond to a resolution in range coding
frequency of 2 Hz, and a resolution rate of change of the coding
frequency equal to 0.2 Hz/s. No physical display system has shown
to provide such high resolution. Simmons (1971, 1973) [31,32]
experimented with the FM bats, and found that the two species

Eptesicus and Phyllotomus could have an acuity in discriminating
target range of 1 to 2% at the absolute distances of 60 cm and
120 cm. One specimen of Eptesicus was found to be able to
discriminate two targets of 14 mm apart at the absolute distance
of 240 cm, i.e. $\Delta r/r \doteq 0.6\%$. It has been proposed that these species
have an ideal sonar receiver which cross correlates the transmitted
and received signals to extract target-range information. The
correlation process may take place at the higher level of
perception than the basilar membrane of the bat. This discovery
would be an impetus for researchers to look for a real time display
of high resolution and fast response for the CTFM sonar which could
emulate the performance of the bat.

REFERENCES

1.   Anke, Von D. "Blindenorientierungshielfen mit
     ultraschallortungs Verfahren und Horbarer Anzeige", Acustica,
     Vol.30, pp.67-80, 1974.

2.   Arasian, P. "Evaluation of the Binaural Sensory Aid for the
     Blind", A.F.B. Res. Bull., Vol.26, pp.35, 1973.

3.   Bainbridge, R. "The Speed of Swimming of Fish as Related to
     Size and to the Frequency and Amplitude of the Tail Beat",
     J.Exp. Biol., Vol.35, pp.109-133, 1958.

4.   Do, M.A. and Kay, L. "Resolution in an Artificially Generated
     Multiple Object Auditory Space Using New Auditory Sensations",
     Acustica, Vol.36, pp.9-15, 1976/77.

5.   Do, M.A. "Perception of Spatial Information in a Multiple
     Object Auditory Space", Ph.D. Thesis, University of Canterbury,
     1977.

6.   Fletcher, H. "Auditory Patterns", Review of Modern Physics,
     Vol.12, pp.47-65, 1940.

7.   Fletcher, H. "Speech and Hearing in Communication",
     D. Van Nostrand Co. Inc., New York, 1953.

8.   Greenwood, D.D. "Auditory Masking and the Critical Band",
     J. Acoust. Soc. Am., Vol.33, pp.248, 1961.

9.   Goldstein, J.L. "Auditory Nonlinearity", J.Acoust. Soc. Am.,
     Vol.41, pp.676-688, 1967.

10. Hall, J.L. "Auditory Distortion Products, f_2-f_1, and $2f_1-f_2$",
     J.Acoust. Soc. Am.,Vol.51, pp.1863-1871,1972.

11. Hall, J.L. "Monaural Phase Effect: Cancellation and
     Reinforcement of Distortion Products f_2-f_1 and $2f_1$-f_2",J.Acoust.
     Soc.Am., Vol.51, pp.1872-1881, 1972.

12. Hamilton, P.M. "Noise Masked Threshold on a Function of Tonal
     Duration and Masking Noise Bandwidth", J.Acoust. Soc. Am.,
     Vol.29, pp.506-511, 1957.

13. Harden Jones, F.R. "Fish Migration", Edward Arnold Publishers
     Ltd., London, pp.230-234, 1967.

14. Hawkins, J.E. and Steven, S.S. "The Masking of Pure Tone and
     Speech by White Noise", J.Acoust. Soc. Am., Vol.22, pp.6-13,
     1950.

15. Hester, F.J. "Identification of Biological Sonar Targets
     from Body Motion Doppler Shifts", Symposium on Marine
     Bio-Acoustics, Pergamon Press, New York, Vol.2, pp.59-74, 1967.

16. Kay, L. "A New or Improved Apparatus for Furnishing
     Information as to Position of Objects", Patent Specification
     No.978741, The Patent Office, London, 1959.

17. Kay, L. "Blind Aid", Patent Specification No.3366922, United
     States Patent Office, Washington, D.C., 1965.

18. Kay, L. "A Sonar Aid to Enhance Spatial Perception of the
     Blind: Engineering Design and Evaluation", Radio and Electron.
     Eng., Vol.44, pp.605, 1974.

19. Kay, L. "Sonic Glasses for the Blind - Presentation of
     Evaluation Data", A.F.B. Res.Bull., Vol.26, pp.35, 1973.

20. Kay, L. and Do, M.A. "An Artificially Generated Multiple
     Object Auditory Space for Use where Vision is Impaired",
     Acustica, Vol.36, pp.1-8, 1976/77.

21. Lichte, W.H. and Gray, R.F. "The Influence of Overtone
     Structure on the Pitch of Complex Tones", J. Exp.Psychol.,
     Vol.49, pp.431-436, 1955.

22. Licklider, J.C.R. "Auditory Frequency Analysis", Symposium
     on Information Theory, The Royal Institution, London, 1955,
     Butterworths Scientific Publications, London, 1956.

23. Madison, D.M. et al "Migratory Movements of Adult Sockeye
     Salmon (Oncorhyn Chus Verka) in Coastal British Columbia as
     revealed by Ultrasonic Tracking", J.Fish. Res.B. Can.,
     Vol.29, pp.1025-1033, 1972.

24. Martin, G. "Electronics and Transducers for an Ultrasonic
     Blind Mobility Aid", M.E. Thesis, University of Canterbury,
     1969.

25. Plomb, R. "The Ear as a Frequency Analyser", J.Acoustic. Soc.
     Am., Vol.36, pp.1628-1636, 1964.

26. Plomb, R. and Levelt, W.J.M. "Tonal Consonance and Critical
     Bandwidth", J. Acoust. Soc.Am., Vol.38, pp.548-559, 1965.

27. Plomb, R. and Steeneken, H.J.M. "Interference Between Two
     Simple Tones", J. Acoust. Soc. Am., Vol.43, pp.883-884, 1968.

28. Rowell, D. "Auditory Display of Spatial Information", Ph.D.
     Thesis, University of Canterbury, 1970.

29. Scharf, B. "Critical Bands and the Loudness of Complex Sounds
     Near Threshold", J.Acoust.Soc.Am., Vol.31, pp.365-390, 1959.

30. Scharf, B. "Loudness of Complex Sounds as a Function of the
     Number of Components", J.Acoust.Soc.Am., Vol.31, pp.783-785,
     1959.

31. Simmons, James A. "Echolocation in Bats: Signal Processing
     of Echoes for Target Range", Science, Vol.171, pp.925-928, 1971.

32. Simmons, James A. "The resolution of Target Range by Echo-
     locating Bats", J.Acoust.Soc.Am., Vol.54, pp.157-173, 1973.

33. Smith, R.P. and Kay, L. "A Fish Finding Sonar Utilizing an
     Audio Information Display", Digest of Technical Papers,
     I.E.E.E., Ocean Conf., Panama City, Florida, pp.113, 1970.

34. Smith, R.P. "Transduction and Audible Displays for Broad
     Band Sonar Systems", Ph.D. Thesis, University of Canterbury,
     1973.

35. Steven, S.S. "Pitch Discrimination, Mels and Koch's
     Contention", J.Acoust.Soc.Am., Vol.26, pp.1075-1077, 1974.

36. Suga, N. "Analysis of Frequency Modulated Sounds by Auditory
     Neurones of Echolocating Bats", J.Physiology, Vol.179,
     pp.26-53, 1965.

37. Suga, N. "Responses of Critical Auditory Neurones to
     Frequency Modulated Sounds in Echolocating Bats", Nature,
     London, Vol.206, pp.890-891, 1965.

38. Suga, N. "Analysis of Frequency Modulated and Complex Sounds
     by Single Auditory Neurones of Bats", J. Physiology, Vol.198,
     pp.51-80, 1968.

39. Swets, J.A. et al "On the Width of Critical Bands", J.Acoust. Soc. Am., Vol.34, pp.108-113, 1962.

40. Von Békésy, G. "Experiments in Hearing", McGraw Hill, New York, 1960.

41. Wegel, R.L. and Lane, C.F. "The Auditory Masking of One Pure Tone by Another and its Probable Relation to the Dynamics of the Inner Ear", Phys. Rev., Vol.23, pp.266-285, 1924.

42. Wever, E.G. "Beats and Related Phenomena Resulting from the Simultaneous Sounding of Two Tones-I", Psychological Review, Vol.36, pp.402-418, 1929.

43. Wever, E.G. "Beats and Related Phenomena Resulting from the Simultaneous Sounding of Two Tones-II", Psychological Review, Vol.36, pp.512-523, 1929.

44. Wever, E.G. "Theory of Hearing", John Wiley, New York, 1949.

45. Zwicker, E. et al "Critical Band-width in Loudness Summation", J. Acoust. Soc. Am., Vol.29, pp.548-577, 1957.

46. Zwicker, E. "Subdivision of the Audible Frequency Range into Critical Bands", J.Acoust. Soc. Am., Vol.33, pp.248, 1961.

Chapter VII
Posters

THE ROLE OF CRANIAL STRUCTURES IN ODONTOCETE SONAR SIGNAL EMISSION

Gustavo Alcuri

Two independent hypotheses attempt to explain the sonar signal transmission mecanism on its path towards the outside environment: 1) focalization through reflection on the concave surface of the anterior cranial region, 2) transmission through the premaxilla.

In order to explain the sonar signal conduction system experimentally, thirty skulls were studied in 23 species from 7 different families, using two fundamental methods, deduction and direct visualization.

The deductive method was applied in several different ways:

1. Anterior cranial region analyses were performed from a geometric point of view. As the configuration of the surface determines the mode of reflection for a given source, the purpose of these analyses was to determine the focalization possibilities of the skull through reflection on the surface of the anterior cranial region.

2. A study was also made of the bone response to vibrations under pulse and sinosoidal excitation. Its purpose was to determine the signal transmission properties of the rostral bone structures. Electrical excitation was transmitted to the bone through a special piezoelectric ceramic whose sandwich-like design allowed simultaneous control of the attack signal. An effective comparison of emitted and received signals could therefore be made after their passage through the biological structure.

3. The propagation velocity of longitudinal and surface waves through the rostral bones was calculated.

4. Attenuation within normal sonar system frequencies was also studied.

Direct visualization of bone structure vibratory behavior was obtained using double exposure holographic interferometry (Cherbit and Alcuri, 1978). This technique permits interference in coherent light of diffraction spectra from excited and non-excited bone. Localized interference fringes can thus be observed on the bone which correspond to movements of the excited structure. The emitting ceramic is placed in the antero-posterior palatine cavity.

   The results using these different methods lead to the following assertions:

1.   Theorectically, the geometrical properties of the skull cannot
explain focalization by reflection in all species.

2.   Signal propagation along the rostrum is responsible for the
significant reduction in the spectral bandwidth of the attack signal
(Fig. 1).

3.   The propagation velocity of longitudinal waves is not constant
along the rostrum but this phenomenon is not connected to the con-
stitutional differences in the bones.

4.   Acoustic vibrations are highly attenuated during their propagation through the bone in the rostrum considered globally (Fig. 1).

5.   From a vibratory point of view, the superior maxillary bones
respresent a functional identity.   In other words, the different
rostral bones act together as a whole (Fig. 1 and 2).

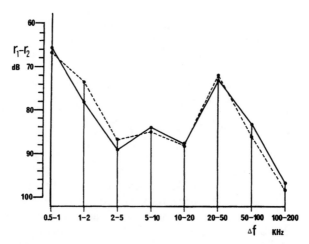

Fig. 1.   Frequency interval/reference level- response level $(r_1 - r_2)$
         in D. delphis.

         ———————   Premaxillary

         - - -   Maxillary

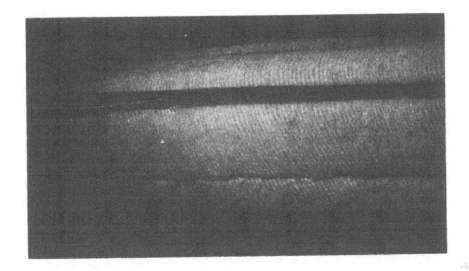

Fig. 2. Direct visualization of bone vibratory behavior by double
         exposure holographic interferometry. The result of this
         is an equivalence between the left and right sides of the
         rostrum and a functional continuity between the premaxilla
         and the maxilla (Sotalia teuzii).

         Excitation frequency: 20 kHz.

Echolocation signal transmission by internal bone conduction
therefore seems unlikely. The results indicate that signal propagation takes place through the interface constituted by the premaxilla and the cephalic soft tissue. In some groups these transmission waves are complementary with the reflection method.

Cherbit, G. and Alcuri, G., 1978, Etude de la propagation des
         vibrations à travers le rostre de Sotalia teuzii (Cetacea)
         par interférométrie holographique.C.R. Acad. Sc. Paris,
         286: D.

BINAURAL ESTIMATION OF CROSS-RANGE VELOCITY AND OPTIMUM

ESCAPE MANEUVERS BY MOTHS

Richard A. Altes and Gerald M. Anderson

Roeder (1970) has shown that noctuid moths display erratic flight patterns during pursuit by bats, and that these moths are capable of sensing their approximate distance from the bat (from the loudness of echolocation signals) and the direction of the bat (from comparison of two sound sensors on opposite sides of the body). This behavior suggests the following two problems: (1) What is the best maneuver the moth can make?, and (2) Is the bat theoretically capable of sensing such a maneuver?

Problem (1) was solved by an application of zero-sum, perfect information, differential game theory (Isaacs, 1967). Assuming that the bat has perfect information about the moth's position and velocity, the theory yields a berrier which separates the bat's capture region (assuming the moth maneuvers optimally). From the data of Webster (1966), it would appear that a bat's velocity is ≈5 m/s, a moth's velocity is ≈1.5 m/s, a bat has a turn radius of ≈1 m, a moth has a turn radius of ≈0 m, and the bat can capture anything within a circle of diameter 5 cm (e.g., with a somersault maneuver). Using these parameters, the capture region has a minimum radius at a range of ≈30 cm from the bat, or ≈.06 sec before capture. The moth's best maneuver is to fly normal to the bat's velocity vector, and the best chance for escape occurs ≈30 cm from the bat.

Problem (2) was solved by assuming that a bat can measure a differential Doppler shift between the two ears. Using this differential Doppler shift, the standard deviation of an optimum estimate of tangential velocity v_t in the horizontal plane is:

$$\Delta v_t = rc/(\sqrt{2SNR}\ Tf_0 d)$$

for a CW bat at range r with signal frequency f_0, pulse length T, and interaural distance d. If the bat first detects its prey at r = 3 m, and if the SNR at this distance is 3 dB, then at r = 30 cm, SNR is at least 2×10^4, and $\Delta v_t = (32T \times 10^2)^{-1}$ when c = 330 m/s, f_0 = 80 kHz, and d = 2 cm. Typically, a CW bat pulse shortens to roughly T = 2 msec at 30 cm, but even with such a short pulse, $\Delta v_t \approx 0.2$ m/s. The bat's theoretical resolution capability is thus sufficient to detect an optimum escape maneuver with a horizontal velocity component greater than 0.2 m/s, near the moth's best escape position. If only interaural frequency shift were used to estimate cross-range velocity, then the moth's best strategy would be a dive in the bat's median plane, normal to the bat's velocity vector.

DOLPHIN WHISTLES AS VELOCITY-SENSITIVE SONAR/NAVIGATION SIGNALS

Richard Altes and Sam H. Ridgway

Certain dolphin whistles display a hyperbolic increase in frequency as a function of time, followed by a hyperbolic decrease. These whistles nearly always contain one or more harmonics. The signals were classified as distress calls by Lilly (1963), but they are also used under many other circumstances.

The above dolphin whistles are very similar to signals that can be used for accurate Doppler measurement (Altes and Skinner, 1977). Velocity is measured by observing the Doppler-induced phase shift in a signal $a(t) \exp(j \cdot b \cdot \log t)$. A Doppler-scaled echo for such a signal is:

$$s^{1/2} a\{s(t-\tau)\} \exp\{jb \log s(t-\tau)\} \simeq a(t-\tau) \exp\{jb \log(t-\tau) + jb \log s\}$$

where s is a velocity-dependent scale factor and τ is delay.

Objections to this theoretical conjecture are as follows:

1. Mammals are insensitive to a constant phase shift at high frequencies.

2. The actual dolphin whistles are often asymmetric in time, instead of symmetric, as specified by Altes and Skinner.

3. The whistles sometimes display rapid fluctuations in frequency and amplitude, relative to the theoretical versions.

These objections are satisfied by the following arguments:

1. Phase insensitivity in mammalian audition does not apply to relative phase shifts between signals at two different frequencies, and the relative phase shift of a signal and its harmonic, $2b \log s - b \log s$, should be audible.

2. The formulation in Altes and Skinner can be generalized to yield asymmetric signals with no range-Doppler coupling.

3. Rapid, zero-mean frequency and amplitude fluctuations
    are "averaged out" by the velocity estimator, and
    have little effect upon the estimates.

The next step is to perform behavioral experiments to determine the jnd in scale factor for scaled, linear-period-modulated
signals with harmonics. Are these minimum perceptible scale differences small enough to be useful for sonar/navigation?

REFERENCES

Altes, R. A., and Skinner, D. P., 1977, Sonar velocity resolution
    with a linear-period-modulated pulse, J. Acoust. Soc. Amer.,
    61:1019.
Lilly, J. C., 1963, Distress call of the bottlenose dolphin:
    stimuli and evoked behavioral responses, Science, 139:116.

TARGET RECOGNITION VIA ECHOLOCATION BY <u>TURSIOPS</u> <u>TRUNCATUS</u>

Whitlow W. L. Au and Clifford E. Hammer, Jr.

Experiments were conducted to investigate the target echo recognition and discrimination behavior of <u>Tursiops</u> <u>truncatus</u>. Two specific hollow aluminum cylinders (3.81 cm and 7.62 cm O. D.) and two solid coral rock (3.81 cm and 7.62 cm diameter) cylinders, all 17.8 cm in length, were used as standard baseline targets. The coral rock targets were constructed of coral pebbles encapsulated in a degassed epoxy mix. Illustrated in Fig. 1 is the experimental pen with the two target funnels located at ranges of 6 and 16 m in front of the dolphin's pen. The funnels were used to present the targets into the water at approximately the same location at each range. The two-alternative, forced-choice technique was used with the targets presented successively in a random sequence. The animal was required to echolocate on the targets and respond to paddle A for the aluminum and paddle B for the coral rock standards. After near errorless performance with the standard targets was achieved, probe sessions were conducted to investigate differential reporting of probe targets varying in structure and composition. All of the probe targets were cylinders of 17.8 cm length. Two probe targets were used in each probe session and only 8 of the 64 trials of the session were used for probe trials, 4 for each probe target.

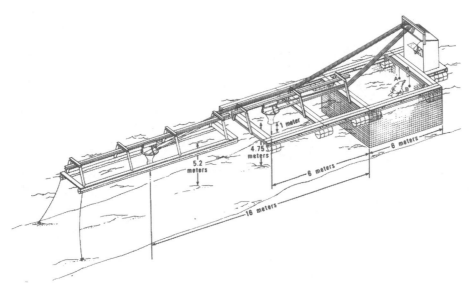

Fig. 1. Experimental Configuration.

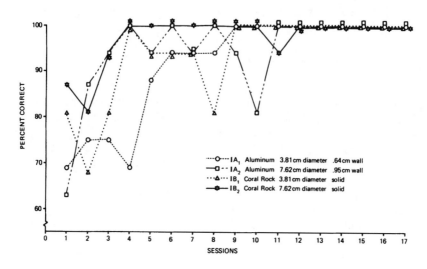

Fig. 2. Acquisition curves for the baseline targets.

RESULTS

The initial acquisition of the baseline standard targets are shown in Figure 2. The dolphin reached asymptotic performance with near errorless behavior after session 11. For experiment I, a general discrimination test was made with one rock, one PVC, two solid and two hollow aluminum, and two corprene cylinders as probe targets. Initially the dolphin classified 3 of the aluminum probes with the standard aluminum targets but eventually classified all of the probes with the coral rock standards, after the target range was moved from 6 to 16 m and then back to 6 m. The dolphin began to focus in on the salient features of the two aluminum standards and reject all other targets as being "not A". In other words, the dolphin became a null detector and searched for the presence or absence of the aluminum standards. In order to further test the dolphin's discrimination capability, hollow aluminum probes with the same outer diameters but different wall thicknesses as the aluminum standards were used in experiment II. The results showed that the dolphin could reliably discriminate wall thickness difference of 0.16 cm for the 3.81 cm 0. D. targets, and 0.32 cm for the 7.62 cm 0. D. targets. In experiment III, the dolphin's ability to discriminate material composition was tested using bronze, steel and glass probes that had the same dimensions as the aluminum standards. It was found that the animal could discriminate the bronze and steel targets from the aluminum but classified the glass probes with the aluminum standards.

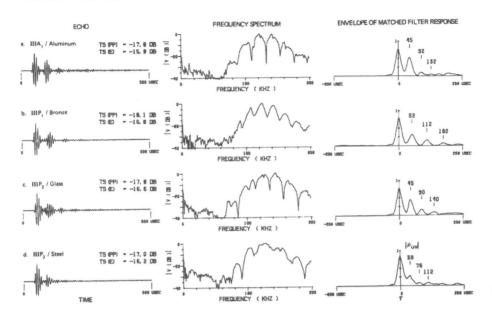

Fig. 3.   Acoustic measurement results.

DISCUSSION

In order to determine the acoustic cues used by the dolphin,
all of the targets were examined acoustically, using a simulated
dolphin echolocation signal. The results for the small cylinders
used in the material composition experiment are shown in Fig. 3.
The envelopes of the matched filter response with the reference
signal being the outgoing signal, are included in the figure. The

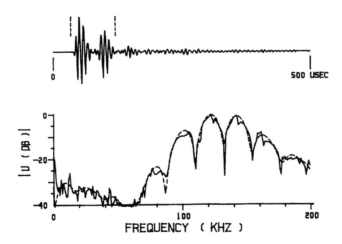

Fig. 4.   Frequency spectra for the small aluminum standard.

matched filter model indicated that the arrival times of the secondary echo components for the aluminum and glass cylinders were very
similar, whereas the arrival times for the bronze and steel cylinders were quite different than for the aluminum. This indicates
that time of arrival cues may have been the primary cues used by
the animal in discriminating the targets. The arrival time difference between the first and second echo components probably resulted
in a time separation pitch, since these two echo components are
highly correlated. The spectrum for the first two echo components
is shown in Fig. 4, overlapping the total epectrum for the small
aluminum standard. Note how well the total spectrum is described
by the spectrum for the first two echo components. Furthermore,
such a rippled spectrum will introduce a time separation pitch that
should be perceived by the dolphin. The echoes were also analyzed
with a contiguous constant-Q filter bank of ten filters using a Q
of 13.8 to simulate the dolphin's auditory system. The results of
this analysis indicated that the bronze probe was more similar to
the aluminum than was the glass probe. However, the dolphin's performance indicated otherwise.

SPHERE-CYLINDER DISCRIMINATION VIA ECHOLOCATION

BY TURSIOPS TRUNCATUS

Whitlow W. L. Au, Ronald J. Schusterman,

and Deborah A. Kersting

An experiment was conducted to investigate the ability of
T. truncatus to discriminate between spherical and cylindrical targets by echolocation. The targets were constructed out of poly-
urethane foam to eliminate any internal reflections, and their dimensions were chosen so that the spheres and cylinders had overlapping target strengths. All targets were acoustically examined with
dolphin-like echolocation signals. The results of the acoustic
measurements performed under free-field conditions are shown in
Table 1, along with the correlation coefficients between the echoes.

Two spheres and two cylinders were used in each 64 trial session except during the first portion of the initial acquisition
phase (see first panel of Fig. 1). The targets were placed on an
assembly which allowed them to be submerged (1.12 m depth) at approximately the same location; 6 meters in front of the animal's
pen. The two-alternative forced-choice technique was used with the
targets presented successively in a random sequence. During the
initial acquisition phase, the dolphin was allowed to swim freely
while echolocating. Eventually, the animal was required to station

Table 1. Results of the Acoustic Examination of the Targets

	Target	Target Strength Measurements			Correlation Coefficients Between Target Echoes		
		Diameter	Length	TS	S_1	S_2	S_3
Spheres	S_1	10.2 cm	–	–32.1 dB	1.00	0.98	0.99
	S_2	12.7 cm	–	–31.2 dB	0.98	1.00	0.99
	S_3	15.2 cm	–	–28.7 dB	0.99	0.99	1.00
Cylinders	C_1	1.9 cm	4.9 cm	–31.4 dB	0.94	0.97	0.96
	C_2	2.5 cm	3.8 cm	–32.3 dB	0.98	0.97	0.98
	C_3	2.5 cm	5.1 cm	–28.7 dB	0.91	0.93	0.92
	C_4	3.8 cm	3.8 cm	–30.1 dB	0.95	0.97	0.98
	C_5	3.8 cm	5.1 cm	–27.6 dB	0.98	0.98	0.99

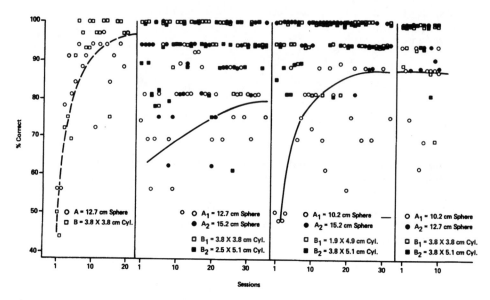

Fig. 1. Initial acquisition results with free-swimming animal.

in a hoop and only echolocate upon the introduction of an audio cue (see Schusterman, Kersting and Au, this volume). The hoop station was introduced to prevent the animal from receiving spatial cues by echolocating at different depths and thus respond on the basis of the aspect dependence of target strength for the cylindrical targets.

Results: The results of the four parts of the initial acquisition phase are shown in Fig. 1. The lines in the figure are curves fitted

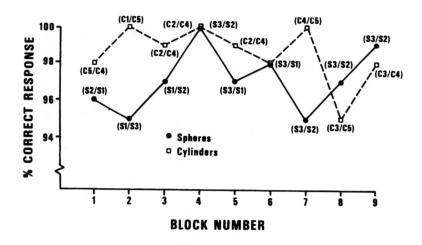

Fig. 2. Summary of the dolphin's in-hoop performance.

to the results of the smallest sphere. The curves show early acquisition with decrements in performance and reacquisition as a function of introducing different cylinders and spheres. Asymptotic
performance was considered to have been achieved as shown in the
last panel of Fig. 1. The results of the dolphin's performance
with the use of the hoop station are illustrated in Fig. 2. Each
block consisted of at least 5 sessions and at most 36 sessions.
A comparison between Fig. 1 and 2 indicates that the dolphin's
ability to differentiate between spheres and cylinders was superior
for the in-hoop situation than for the free-swimming situation.

Discussion: The acoustic examination of the targets showed that
the echoes were very similar, as can be seen in the high correlation
coefficient values of Table 1. It appears unlikely that the dolphin
made the discrimination based on differences in the measured reflec-
tive characteristics of the targets made in the free field. Further
acoustic measurements were made in which the planar transducer was
swept in azimuth to simulate the animal scanning across the targets.
These measurements did not provide any information as to how the
animal was making its discrimination. Finally, the transducer-target

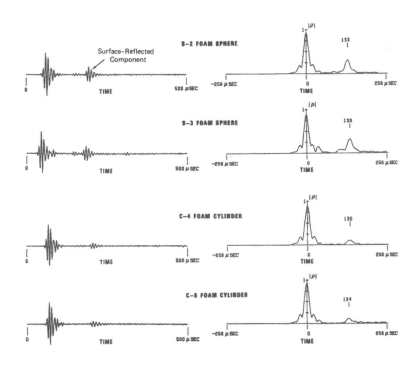

Fig. 3. Approximation of echoes received by the dolphin.

geometry was set up to approximate the animal's experimental condition. Examples of the results of this acoustic measurement are shown in Fig. 3. Also shown in the figure are the envelopes of the matched-filter responses, with the filter matched to the outgoing signal. It was found that the spheres introduced a larger surface reflected component to the echoes than did the cylinders. Furthermore, the surface reflected component of the cylinder echoes fluctuated in amplitude as wind and wave motion caused movements in the targets. The surface reflected component of the echoes seemed to have provided the eventual cue for the dolphin's differentiation between the targets. Superior performance in the hoop station over the free-swimming situation is consistent with the notion that the dolphin used the surface reflected component of the echoes as the primary cue.

FREQUENCY SELECTIVITY OF CONSTANT LATENCY NEURONS IN THE INFERIOR

COLLICULUS OF MEXICAN FREE-TAILED BATS

Robert Bodenhamer, George Pollak, and David Marsh

There are neurons in the inferior colliculus of the Mexican free-tailed bat, Tadarida brasiliensis mexicana, which are characterized by a highly stable discharge latency to frequency modulated (FM) signals. These phasic constant latency responders (pCLRs) are capable of accurately encoding the time interval between simulated orientation cries and echoes, and thus appear to be well suited for encoding target range.

Each pCLR is triggered by a particular frequency component of a downward sweeping FM burst. This can be shown by presenting bats with FM signals having the same modulation depth (typically sweeping from 50 kHz to 25 kHz, mimicking the orientation cries of Tadarida), but having different durations (Fig. 1). For each pCLR, the discharge latency increases as the signal duration is lengthened. The difference in latency from one duration to another consistently corresponds to the shift in the temporal position of a particular frequency component within the signal. Since a pCLR "follows", or is excited by, this component, we have labelled it the cell's excitatory frequency (EF). To date, units have been found with EFs ranging from 24.5 kHz up to 46 kHz.

The pCLRs' tight synchronization to very limited frequency ranges within a modulated signal might suggest that they are more sharply tuned to constant frequency signals than are other collicular neurons. That this is not the case is demonstrated by the representative tuning curves shown in Fig. 2. As a rule, tuning curves of pCLRs and of other collicular cells are indistinguishable with Q-10 dB values of from 3-11 being most common for each group.

Sensitivity to a particular EF represents a high degree of frequency selectivity, rivaling that of the sharply tuned filter neurons of the long–CF/FM bats such as Rhinolophus ferrumequinum. However, in Tadarida, this selectivity is apparently not due to mechanical specialization of the cochlea, but rather to complex neural circuitry.

In conclusion, the pCLRs' sensitivity to a particular EF is valuable to FM bats in at least two ways. First, it leads to the tight firing registration enabling the pCLRs to serve as pulse-echo time markers, and thereby encode target range information. Second, by restricting the response area of each pCLR to a small frequency band, it may enable a population of pCLRs to encode echo spectral

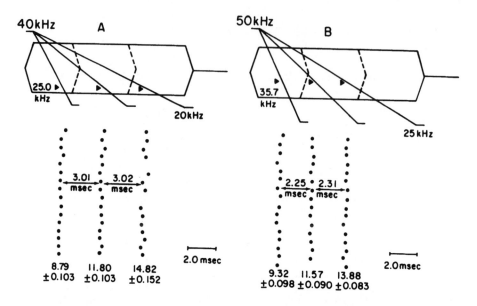

Fig. 1. Responses of two pCLRs which illustrate sensitivity to a
         particular EF. Top portion of both A and B show FM pulses
         of three durations. EAch dot column represents action
         potentials elicited by 16 stimulus presentations at a
         given duration. The first column in each half of the
         figure was produced in response to the 4.0 msec dura-
         tion signal, the second to the 8.0 msec duration sig-
         nal, and the third to the 12.0 msec duration signal.
         The dot columns have been aligned directly beneath the
         FM signals to show how they line up with the temporal
         position of the proposed EF in each sweep (indicated
         by arrows). Each dot column's actual mean response
         latency and standard deviation of mean latency is shown
         below that dot column (in msec). In A, the FM signals
         sweep from 40 kHz to 20 kHz and the calculated EF is
         25.0 kHz. Unit T10-12-88, 77 dB SPL. In B, the sig-
         nals sweep from 50 kHz to 25 kHz and the calculated EF
         is 35.7 kHz. Unit T82-5, 58 dB SPL.

characteristics, and thus to convey information about the physical
nature of targets as well as target range.

(Supported by NIH Research Grant NS-13276.)

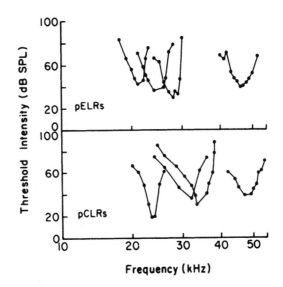

Fig. 2. Representative tuning curves of pCLRs (below) and
of other units for which response latency was not
constant (above). The important feature is the
similarity in the Q-10 dB values of the two groups.

STRUCTURAL ADAPTATION IN THE COCHLEA OF THE HORSESHOE BAT FOR THE

ANALYSIS OF LONG CF-FM ECHOLOCATING SIGNALS

Volkmar Bruns

The auditory system of the greater horseshoe bat, Rhinolophus ferrumequinum, is sharply tuned to the echo of the constant-frequency (CF) segment of the orientation signal of about 83 kHz (Neuweiler, 1970). The tuning is performed within the cochlea and is correlated with a number of specialized features (Bruns, 1976a and 1976b; Schnitzler et al., 1976; Suga et al., 1976).

The functional morphology of the cochlea of the horseshoe bat, studied last years in our Frankfurt group, is integrated in the review of G. Neuweiler, in this volume. This paper presents a mechanism creating the sharp cochlear tuning around 83 kHz by a transversal vibration component of the basilar membrane (BM) and its associated structures which is superimposed onto the longitudinal travelling wave.

The specialized basal region of the horseshoe bat cochlea where the CF-component is analyzed is shown in Fig. 1,B. In comparison to the cochlea of other "high frequency" mammals adapted to the analysis of a broad frequency band (e.g., FM-bats; see Fig. 1,A) the horseshoe bat demonstrated following specializations: (1) A displacement of the feet of the outer pillar from the thin center of the BM onto the inner thickening resulting in a pliable connection between the inner and outer segment of the BM. (2) The spiral lamina, with the exception of its inner margin, is almost free of radial fibers. Between the thickening of the secondary spiral lamina and the outer bony wall there is a small bone lamella. Thus the BM is only loosely attached to the outer bony wall. (3) The outer segment of the BM is connected with the thickening of the secondary spiral lamina by filaments of the inner margin of the spiral ligament. This part of the spiral ligament contains fibers in a longitudinally oriented bundle. This is interpreted that in between the loose structures in point (1) and (2) a rigid vibration unit exists enclosing the outer thickening of the BM and the secondary spiral lamina with the possibility of rotation around the axis of the longitudinal fibers in the spiral ligament.

In capacitive probe measurements of the vibration behavior of the BM and associated structures (joint studies with J. P. Wilson) two main results are found in the specialized region of the horseshoe bat cochlea. (1) The adjacent structures of the BM, the primary spiral lamina and the inner margin of the spiral ligament vibrate 35 and 20 dB lower than the BM. However, the thickening of

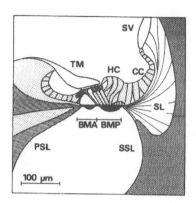

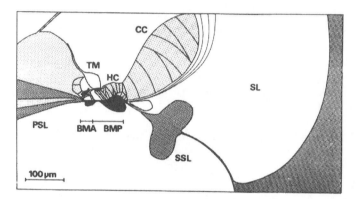

Fig. 1.   Cross sections of the organ of Corti of two high-frequency
          hearing mammals emphasizing the hydro-mechanically import-
          ant structures.   Adaptations for the analysis of (A) a
          broad-band echolocating signal (e.g., FM bats) and (B)
          a small-band echolocating signal (CF bat, Rhinolophus
          ferrumequinum).   BMA: Basilar membrane, arcuate zone;
          BMP: Basilar membrane, pectinate zone;   C: Tunnel of
          Corti;   CC: Claudius' cells;   HC: Hensen's cells:
          PSL: Primary spiral lamina;   SSL: Secondary spiral lamina;
          SL: Spiral ligament;   SV: Stria vascularis;   TM: Tector-
          ial membrane.

the secondary spiral lamina (see Fig. 1,B) vibrates with an unexpected high amplitude, only 5 dB lower than the BM. (2) The inner
and outer segments of the BM are in phase up to about 80 kHz, then
diverge and pass through an antiphase at the location of 83 kHz,
beyond which the phase difference increases further until it reaches
one full cycle for about 86 kHz.

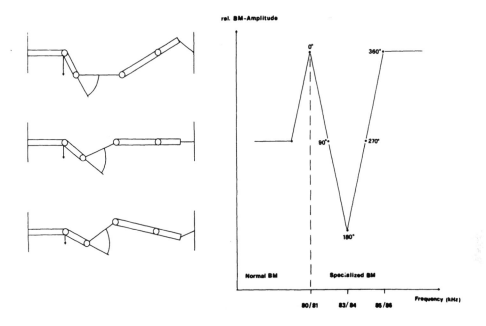

Fig. 2. (A) Model of the basilar membrane and associated struc-
         tures of the horseshoe bat at different phase relations.
         (B) The curve of mechanical frequency sharpening.

        The fine structure and the vibration behavior are integrated
in a model of mechanical frequency sharpening (Fig. 2,A). In the
vibration modes of Fig. 2,A the angle β, a measure for the transversal vibration component of the receptor site of the organ of
Corti, is constant. The amount of vertical BM-amplitude (arrow)
required to achieve in the same angle depends on the phase relations
between the inner and outer vibrating units (double lined). In
phase (above) a high BM-amplitude is needed. The amplitude at 90°
phase divergence (middle) is medium and small at 180° (antiphase;
below). 270° corresponds to 90°, and 360° is in phase again. The
amplitude of the normal BM, in the cochlear region below 80 kHz,
is similar to the middle mode in the specialized BM. The curve
(Fig. 2,B) in which the relative BM-amplitude of the model in Fig.
2,A is correlated with the frequencies, fairly well matches to those
of the neuronal audiogram, around the frequency of the CF-component
of the echolocating signal.

SIMILARITIES IN DESIGN FEATURES OF ORIENTATION SOUNDS USED BY SIMPLER, NONAQUATIC ECHOLOCATORS

Edward R. Buchler and Andrew R. Mitz

There are some remarkable similarities in the echolocation pulses of several species of nonaquatic echolocators that are relatively simple or conservative in terms of their echolocating abilities. I would like to illustrate this and offer speculations as to why this particular pattern of sounds has evolved repeatedly and, in many cases, independently. In the primary examples given, the echolocation signals consist of brief, unstructured sounds having a rapid onset and broad bandwidth. In addition, they often or always occur as paired pulses.

Supporting Data. Recently, Thurow and Gould (1977) presented strong circumstantial evidence that the broadband pulse pairs of a caecilian (probably Dermophis septentrionalis) are used for orientation. Each sound lasts ca. 20 ms and extends from <500 Hz to >25 kHz. The onsets of members of a pair are separated by ca. 200 ms.

The vagrant shrew, Sorex vagrans, emits paired pulses while echolocating (Buchler, 1976). Each pulse lasts from <1 ms to 2.5 ms. The members of each low intensity pulse pair are ca. 8-10 ms apart and span a frequency range of at least 20-60 kHz (Fig. 1). The broadband ultrasonic clicks of the short-tailed shrew, Blarina brevicauda, also are temporally associated most commonly as pairs (Tomasi, pers. comm.).

Three, and probably four, genera of tenrecs (Tenrecidae) echolocate using tongue clicks of 5-17 kHz which last only 0.1-3.6 ms. Again, the most common temporal configuration is pairs (Gould, 1965).

Chase (this volume) has recorded the very brief, low intensity clicks used by echolocating laboratory rats (Rattus norvegicus). They commonly are emitted in pairs or trains of pairs.

Three of the four most studied species of echolocating cave swiftlets (Aerodramus = Collocalia) use broadband clicks of uncertain origin. The frequencies range from 2-15+ kHz. There are ca. 15 ms between members of a pair (Griffin and Suthers, 1970; Fenton, 1975; Medway and Pye, 1977).

Finally, a few species of the megachiropteran, Rousettus, echolocate by using paired tongue clicks ranging from 10 kHz to as high as 100 kHz. The interclick interval is ca. 20-30 ms, each click being only a few milliseconds long (see Sales and Pye, 1974).

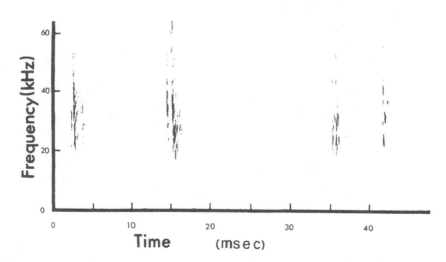

Fig. 1.   Two pairs of sonar pulses emitted by a vagrant shrew, Sorex vagrans.

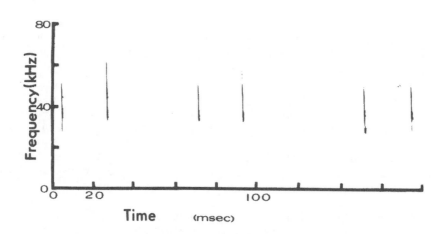

Fig. 2.   Three pairs of FM sweeps emitted during the first flight of a little brown bat, Myotis lucifugus.

In addition, there are other species, some related to those
above, that use broadband clicks that are not commonly paired; e.g.,
some shrews, a cave swiftlet and the oil bird, Steatornis. There
appear to be correlates of this kind of broadband signal in some of
the anatomically more primitive Microchiroptera. Megaderma, Rhinopoma
and Nycteris use sounds of only a few milliseconds which have 3-5
harmonics and virtually no frequency modulation (Simmons, 1979).
There might well also be ontogenetic correlates of the paired pulse
phenomenon in the development of echolocation in some vespertilionids.
The little brown bat, Myotis lucifugus, uses paired cruising and
approach pulses during the first few nights of flight. The FM sweeps
of a given pair are ca. 20 ms apart (Fig. 2) (Buchler, 1979).
Although this pairing is quickly lost as the echolocation system of
M. lucifugus matures, Woolf (1974) believes that a very similar form
of pairing persists even into adulthood in the big brown bat,
Eptesicus fuscus, and has its origin in a double note call. Pairing
also occurs in the early flights of the cave bat, Myotis velifer
(Brown, pers. comm.).

    Discussion. It is obvious that very short pulses and their
related wide bandwidth provide good range resolution and localization characteristics; but why the recurrence of pairing? One important adaptive advantage of multiple pulse sonar is that, by integration of successive similar pulses, the signal to noise ratio is increased. Regardless of the technique the individual uses to scan
its environment, bounds can be set on the effectiveness of its inte-
grating system. Consider a poor, but not worst, case near the lower
bound. For a simple, or primitive, uniformly weighted, noncoherent
integrator:

$$(S/N)_n \approx \sqrt{n}(S/N)_1$$

This illustrates that the integration of multiple pulses has rapidly
diminishing returns associated with it (Skolnik, 1962). A second
pulse increases the S/N by 3 dB, but a third by only 1.8 dB, etc.
Thus, the paired pulse may be an energetically efficient strategy
by which terrestrial insectivores, for example, with low intensity
pulses can increase the S/N ratio while still retaining the advantages of short duration and broad bandwidth. Those species that use
higher intensity pairs and move more quickly, such as the cave swift-
lets, may be able to derive relative velocity information by comparing the interval between members of the echo pair with that of the
emitted pair. The interval will be subject to the Doppler effect
even though a given pulse is Doppler tolerant.

It should be noted that the paired click structure is also used commonly by several cetaceans; e.g., <u>Tursiops</u> (Norris et al., 1967), <u>Sotalia</u> (Norris et al., 1972) and <u>Delphinapterus</u> (Evans, pers. comm.). Although Mackay (1967) suggests that the second click may function to greatly decrease ringing in the sending or receiving apparatus, the explanations offered here may be more generally applicable.

REFERENCES

Buchler, E. R., 1979, The development of flight, foraging and echo-location in the little brown bat, <u>Myotis</u> <u>lucifugus</u>, <u>Behav.</u> <u>Ecol</u>. <u>Sociobiol</u>., (in press).
Simmons, J. A., 1979, Phylogenetic adaptations and the evolution of echolocation in bats (Chiroptera), <u>Proc</u>. <u>5th</u> <u>Internat</u>. <u>Bat</u> <u>Res</u>. <u>Conf</u>., (in press).

RAT ECHOLOCATION: CORRELATIONS BETWEEN OBJECT DETECTION AND CLICK PRODUCTION

Julia Chase

The laboratory rat Rattus norvegicus uses a variety of sensory cues to guide its nocturnal movements. Vision, olfaction, touch, and a well-developed spatial memory all aid the rat in orientation and obstacle avoidance. A half century ago, Shepard (1929), Lashley, (1929), and Honzik (1936) found that hearing is also important for orientation when a rat is deprived of visual cues. These experiments were published a decade before the discovery of echolocation in bats and 25 years before the detection of ultrasonic vocalizations in rats (Anderson, 1954) and therefore aroused little interest.

Riley and Rosenzwieg (Rosenzwieg et al., 1955; Riley and Rosenzwieg, 1955) demonstrated (1) that blind rats on an elevated Y-maze could determine which of two runways was obstructed by a barrier, and (2) that this discrimination broke down when the rat was deafened or alternatively if the barrier was rotated 45°. Bell et al., (1971) confirmed these findings and demonstrated that white noise of 20 dB also abolished the discrimination. Though both studies concluded the rats were echolocating, neither found evidence of sound production other than occasional scratches or sniffling noises. The results are thus open to two interpretations: (1) the rats are actively echolocating but emitting sounds too faint to be detected by the equipment, or (2) the rats are passively listening for differences in the ambient noise levels on the two runways. The published data remain as inconclusive as they were when Neuweiler reviewed the topic at the first of these symposia 13 years ago (Neuweiler, 1966).

Work currently in progress in this lab has confirmed Riley and Rosenzwieg's behavioral findings. With the improved equipment currently available, we were able in addition to detect brief, faint, broadband clicks that were not noted by the earlier investigators. The following progress report presents our results to date which suggest that these clicks are used by the rats as part of an active echolocation system.

Blind, water-deprived Long Evans rats were trained to seek water on an elevated Y-maze. One arm of the Y was open and the other was blocked by a 22 x 27 cm aluminum barrier suspended above the runway at a distance of about 35 cm from the rat's starting position. Sounds were monitored during testing with a Holgate ultrasonic detector tuned to 45 kHz. Preliminary analysis of the sounds was done in D. R. Griffin's laboratory using the more sensitive McCue and Bertolini ultrasound detector (1964) and a LockHeed high speed tape recorder. Taped sounds were photographed with a Grass Camera from a Tektronix oscilloscope.

Eleven rats learned to choose the open runway, scoring 75% or better on 50 consecutive trials. Five of these animals were then ear-plugged with cotton and vaseline and retested. Rats with their ears plugged were unable to select the open runway but when the plugs were removed after 5 days (50 trials) the discrimination was restored.

Detection of the sounds was difficult even with the Holgate and we believe that we were probably not detecting all of the rat's clicks. Nonetheless, there was a clear relationship between the production of clicks and the rat's performance in the maze. First, during training as the rat's performance improved, the rate of sound production increased. Second, while deafening sometimes caused an initial increase in clicks, they soon decreased or stopped altogether. Third, in all eleven animals, the rats were more successful on trials when clicks were detected than on silent trials.

Preliminary analysis of the clicks (Fig. 1.) shows that they are very brief, low intensity, broadband clicks of only 4-8 cycles. In the best recordings made with a carefully adjusted McCue-Bertolini detector, the unfiltered signals are about 3 times the noise level of 35 dB SPL. Filtering out all sounds below 40 kHz and above 50 kHz improved the signal-to-noise ratio.

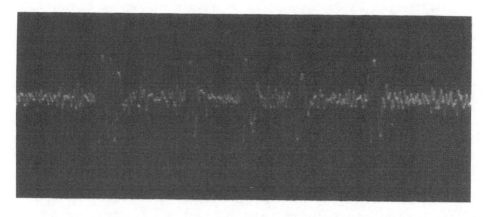

Fig. 1. Short train of clicks produced by a blind rat.
Recorded with an unfiltered McCue-Bertolini
detector and Lockheed tape recorder.

Clicks appeared both singly and in pairs and were sometimes recorded in trains of 30 or more. In these trains, paired clicks comprised about half the sounds produced and were reminiscent of the paired clicks of other echolocators such as shrews, cave swift-lets, tenrecs and Rousettus (see Henson, Buchler, and Mitz, this volume). Clicks also appeared more frequently (or louder?) when the rats were placed in a novel situation such as walking on a yardstick.

ACKNOWLEDGEMENTS

This research was supported by Barnard College and NIMH grant N° 1 R03 MH 30169.

REFERENCES

Bell, R. W., Noble, M. E., and Davies, W. F., 1971, Echolocation in the blinded rat, Percep. Psychophysics, 10:112.
Dunning, D. C., 1975, Orientation by cave-dwelling pack rats, Neotoma floridana magister in the dark, paper presented at the XIV International Ethological Conference, Parma, Italy.

SIGNAL DESIGN FOR MATCHED FILTER DETECTION IN A REVERBERATION-LIMITED ENVIRONMENT: APPLICATION TO CETACEAN ECHOLOCATION SIGNALS

Martine Decouvelaere

This study is concerned with cetacean echolocation signals as optimum signals for Doppler-tolerant, matched filter detection of targets in the presence of reverberation.

A numerical method is derived to find optimum signals in the following situation: a real, fixed-energy signal $s(t)$, $t\varepsilon\{o,T\}$ is emitted. The echo at the receiver is the sum of the known target echo $c(t) = \sqrt{\eta_c} s\{\eta_c(t - \tau_c)\}$, white noise, and spurious echoes from scatterers. The reverberation echo is modeled as a gaussian random process of known expectation and covariance. The covariance is characterized by the response $r(t)$ of an elementary scatterer to $s(t)$ and by the reverberation density function. The receiver is a filter matched to $c(t)$.

The optimum signal, i.e., which leads to a maximum output signal-to-interference ratio, must then satisfy a nonlinear integral equation involving the wideband cross-ambiguity function of $s(t)$ and $r(t)$. This problem can be solved numerically (according to Altes, 1971) by writing $s(t)$ as a linear combination of M orthonormal basis functions of $L^2 \{o,T\}$. Conjectured optimum signals are then found with several choices of reverberation features and basis functions, including Doppler-invariant density functions (in order to yield Doppler-tolerant signals), Rayleigh approximation for $r(t)$, and basis functions derived from Angle Prolate Spheroidal functions.

The signals obtained compare well with experimental cetacean echolocation signals with respect to the above performance criterium as well as Doppler-tolerance and displayed form.

Altes, R. A., 1971, Methods of wideband signal design for radar and sonar systems, Fed. Clearinghouse N° AD 732-494.

VARIATIONS IN THE CHARACTERISTICS OF PULSE EMISSIONS OF A

TURSIOPS TRUNCATUS DURING THE APPROACH PROCESS AND THE ACOUSTIC

IDENTIFICATION OF DIFFERENT POLYGONAL SHAPES

Albin Dziedzic and Gustavo Alcuri

An alternative choice test was used to determine the ability of a 4-year-old Tursiops truncatus to discriminate acoustically a control ring (the positive stimulus) from a regular polygonal frame which was varied between tests. A study was then made of the physical characteristics of the acoustic emissions of the animal during the approach process and the identification of the positive stimulus which the animal was required to bring to the trainer in each trial.

The direct distance between the starting point of the animal and the objects was 8 m. Two acoustical arrays consisting of 10 hydrophones were used to record the signals. These were positioned so as to prevent interference between direct signals of <100 µs duration emitted during approach and their associated first order echoes. Signals were recorded during the complete approach process on a 14 track tape recorder and were synchronized with video recording during the last part of the trajectory (approximately 1.5 m). This allowed us to obtain projections of the trajectory at the end of the approach and to correlate the position of the subject at a specific time with its corresponding acoustic emissions. Discrimination performances are summarized in the following table:

N	3	4	5	6	8	10	12
%	100	100	78	72	77	70	85

where N represents the number of sides of the polygon tested and % the relative number of positive responses obtained with the corresponding polygon. The animal's discrimination performances are better than 70% chance level for all objects.

The video tapes showed three components in the movements of the animal: a) lateral sliding, b) rolling of the body around its longitudinal axis, and c) rotating movements of the head. For certain positions of the animal in its trajectory, an estimation was made of the angular parameters (azimuth θ, site ϕ), as well as of the rolling angle ψ, defining the orientation of the rostrum with respect to the hydrophone. Fig. 1 summarizes the trajectory and associated acoustic data obtained in a test using the pentagon.

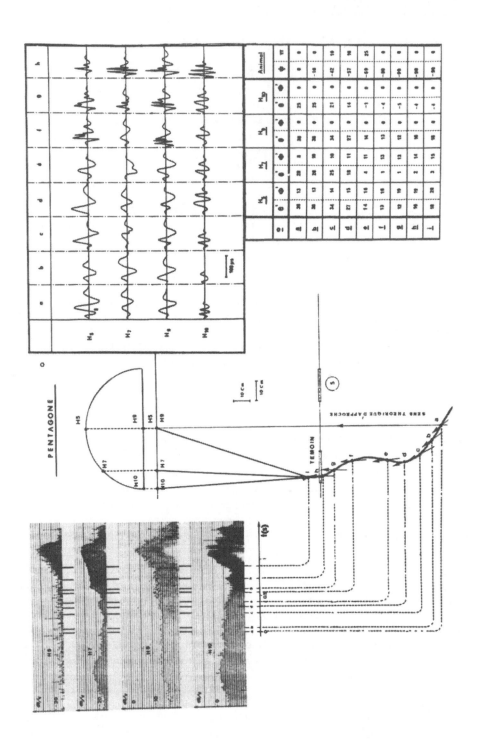

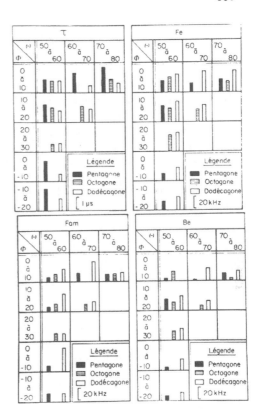

Fig. 2. Results of tests using a pentagon, an octagon, and a dodecagon. The resolution ($\approx 1/\tau_0$) and the other three parameters (F_c, B_e, F_{am}) seem to increase with the number of sides of the polygon.

Fig. 1. A summary of trajectory and associated acoustic data obtained during a test using a pentagon. Only four of the hydrophones are shown (H5, H7, H9, H10).
Upper left: Actographs of the total group of signals emitted during the trial and capted by each of the hydrophones, showing the positions of the animal during selected portions of the trajectory.
Upper right: Oscillographs of signals capted by each hydrophone at these points.
Lower right: Associated angular parameters.

A study was made of the group of signals emitted during each approach and of individual signals emitted at particular positions in the trajectory. For each group of signals, five parameters were considered: 1) the total duration of signal emission, 2) the total number of signals emitted during the trial, 3) the duration of the part of the trial during which the frequency repetition rate (FRR) increased approximately according to $e^{-\alpha t^2}$, 4) the number of signals emitted during this portion of the trial, 5) the maximum frequency repetition rate recorded in the trial. The results indicate that only the number of signals emitted during the first part of the trial, prior to the characteristic increase in FRR, is proportional to the number of sides of the polygon, which is probably related to the difficulty of the discrimination task.

For each signal emitted at a specific point within a trial, four parameters were considered: 1) the first null value of the autocorrelation function, 2) the frequency ν_0 of the spectral density function $\gamma(\nu)$, defined as:

$$F_c = \nu_0 = \frac{\int \nu\gamma(\nu)d\nu}{\int \gamma(\nu)d\nu}$$

3) the equivalent bandwidth of $\gamma(\nu)$, defined as:

$$B_e^{-1} = \frac{\int |\gamma(\nu)|^2 d\nu}{|\int \gamma(\nu)d\nu|^2}$$

4) the dominant frequency F_{am} of $\gamma(\nu)$. Results are shown in Fig. 2.

Signals emitted during roughly the first half of the approach (animal-to-object distance > 4 m) are independent of polygon shape. They are similar for all tests showing high distance resolution and a broadband spectrum. Signals emitted during the second half of the approach (animal-to-object distance < 4 m) differ according to object shape, showing in general lower values for the four parameters than those emitted during the first half.

NEURAL MECHANISMS FOR TARGET RANGING IN AN ECHOLOCATING BAT EPTESICUS FUSCUS

Albert Feng, James A. Simmons, Shelley A. Kick

and Beatrice D. Lawrence

We have concluded a series of behavioral, anatomical and physiological experiments in order to gain a better understanding of the neural basis of target ranging in Eptesicus fuscus. Behaviorally, we were interested in learning whether or not binaural integration is essential for ranging. We trained two bats to perform range discrimination using two-choice paradigms (Simmons, 1973) until they reached a criterion of 85% correct responses. We then observed their performance under intact conditions and when one ear was occluded. Occlusion was effected by inserting silicone grease to fill the ear canal of the selected ear. The performance of these two bats under monaural and binaural conditions is shown in Table 1. We found that the performance under monaural conditions was not significantly different from that of intact conditions ($p < 0.05$, Hoel test for the difference of two means). Thus, bilateral interactions, which are essential for encoding directional information about the target, are not required for ranging in this species. Instead, time or intensity information from target echoes is more important for ranging.

Our physiological recordings from the central auditory structures of Eptesicus demonstrate that pulse-echo time is likely to be the essential information for these bats. We recorded single unit activity from the inferior colliculus, intercollicularis nucleus and the auditory cortex of lightly anesthetized or awake bats, studying the unit responses to paired acoustic stimuli in the free-field simulating pulse-echo sounds that the bat would have heard during prey-catching. We found that a number of neurons in the intercollicularis nucleus and the auditory cortex were selectively responsive to complex acoustic features of the simulated pulse-echo sounds, i.e., they responded only when the sounds were paired and were frequency-modulated, and the interval between the pulse and the echo was not so important in determining the neural responses. These neurons seemed to be "tuned" to a narrow range of pulse-echo delays (Feng, Simmons, and Kick, 1978). Different delay-tuned neurons showed response preferences for varying pulse-echo delays of from 4.5 to 24 msec as shown in Table 2. Thus, target range could be encoded by a place mechanism in the central nervous system as demonstrated most recently in the auditory cortex of Pteronotus parnellii rubiginosus (Suga and O'Neill, this volume).

Table 1.   Correct Responses for Target Range Discrimination

Bat W_2	Number of Trials	% Correct Response
Monaural	51	80%
Binaural	134	78%
Bat G		
Monaural	62	73%
Binaural	286	83%

Table 2.   Response Selectivity to Pulse-Echo Delays

Unit #	Optimal Echo Delay	Response Range
12-1	11 msec	7-15 msec
12-2	9 msec	7.5-12 msec
12-6	10.5 msec	10-12 msec
12-8	12.5 msec	6-16.25 msec
12-9	6.75 msec	4.7-10 msec
13-4	7 msec	4.4-12 msec
13-5	9 msec	5-15 msec
15-6	17 msec	15-20 msec
15-7	13 msec	10-15 msec
15-8	24 msec	20-30 msec

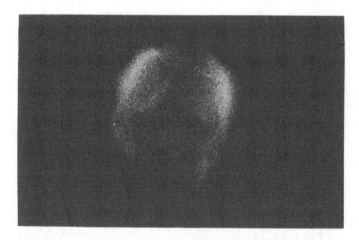

Fig. 1.   Uptake of ^{14}C-2-Deoxyglucose in the intercollicularis
nucleus following stimulation with paired pulse-echo
sound simulating a target at a distance of 2 meters.

The involvement of various neural centers in target ranging was investigated using the labed 2-Deoxyglucose functional mapping technique. In one immobilized bat, we injected about 12 μCi of C^{14}-2-Deoxyglucose intraperitoneally. Subsequently, the bat was exposed to paired acoustic stimuli simulating a target at a distance of 2 meters for 45 minutes. The bat was sacrificed and the brain extracted, frozen, and sectioned. The labeled areas in the brain were observed from X-ray films exposed to sectioned tissues. We found that the inferior colliculus, intercollicularis nucleus (Fig. 1), vast areas of cortex, and lower auditory centers were heavily labeled, indicating their involvement in processing these acoustic stimuli, and perhaps their involvement in processing events related to ranging.

In summary, the results from these combined studies suggest that temporal information may be utilized by Eptesicus for target ranging, and that it is processed along the ascending auditory pathway. Some neurons in the intercollicularis nucleus and the auditory cortex are specialized for processing the temporal delay between pulse and echo and are presumably independent of binaural processing of directional information.

Feng, A. S., Simmons, J. A., and Kick, S. A., 1978, Echo detection and target-ranging neurons in the auditory system of the bat Eptesicus fuscus, Science, 202:645.

Simmons, J.A., 1973, The resolution of target range by echolocating bats, J. Acous. Soc. Amer., 54:157.

Suga, N., and O'Neill, W. E., 1979, Auditory processing of echoes, part II. Representation of acoustic information from the environment in the bat cerebral cortex (this volume).

HUNTING STRATEGIES AND ECHOLOCATING PERFORMANCE OF BATS –

QUANTITATIVE BEHAVIORAL LABORATORY ANALYSIS

J. Fiedler, J. Habersetzer, and B. Vogler

Two bat species, Megaderma lyra and Myotis myotis, were tested behaviorally for the efficiency of their echolocation system. Bats were conditioned to discriminate between two targets, plastic plates with identical pattern of holes of different depths. M. lyra performed 75% correct choices, with hole depth of 8 mm (+) against 6 mm (-). M. myotis succeeded in 87% of flights with depth of 4 mm (+) against 3.2 mm (-). However, the combination of 8 mm and 7.2 mm yields only 68% correct flights. Since the discrimination threshold depends not only on the depth difference, but also on the absolute depths, it is suggested that the bat evaluates small differences in the power spectra of the echo.

How relevant are these performance tests for interpretations of the natural behavior of bats? Both bat species cease to echolocate under certain hunting conditions. In about 46% of 35 trials M. lyra catches living mice without emitting any sound, just by passively listening to the noise generated by the moving prey (Fig. 1a). In the other 54% only sparse echolocation of 3 to 5 ultrasounds per approach is observed. On the other hand, dead mice are detected by means of echolocation (Fig. 1b). This situation is characterized by a typical "paired sound pattern", that was only found in situations where the bat has no other choice but to echolocate.

The echolocation flight of the insectivorous M. myotis consists of 1) search flight, 2) approach, and 3) final buzz. But M. myotis may also switch to passive orientation during hunting on the ground. The bat forages at a distance of 10 to 30 cm without emitting orientation calls. Dead mealworms are located in a 2 cm distance from a wall, and M. myotis crawls to the prey without emitting the final buzz, although the approach sound pattern is present. M. myotis can also rely on passive orientation in flight. It flies to a 6.5 x 6.5 mm mesh within a 40 x 40 cm wire grid if a scratching noise is produced behind the target. It is concluded that M. myotis abandons echolocation when the prey has to be spotted in front of a dense background. The natural prey of M. lyra are also targets on a dense background. For both species it is apparently easier to locate prey on the ground by its own sounds and noises.

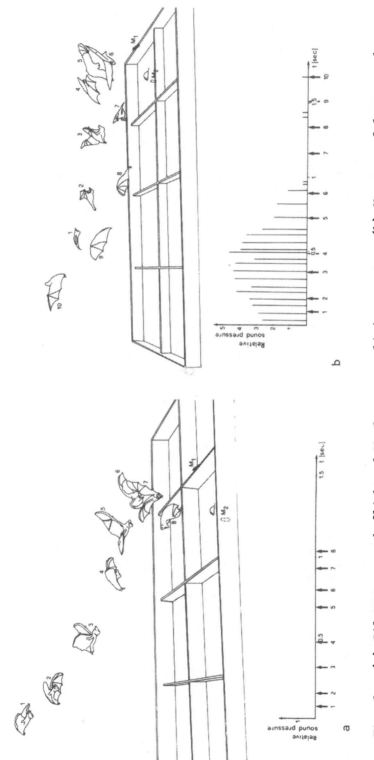

Fig. 1. (a) Silent approach flight of M. lyra to a living mouse. (b) Unsuccessful approach flight of M. lyra towards a dead mouse. M_1 and M_2 indicate the positions of the microphones (M_2 is below the compartment). On the time-axis of the diagram the vertical bars (present only in Fig. 1(b)) indicate echolocation sounds recorded synchronously with the stroboscopic frames. Numbered arrows indicate the corresponding flight position shown above.

LOW-FREQUENCY RECEIVER OF THE MIDDLE EAR IN MYSTICETES AND ODONTOCETES

Gerald Fleischer

Mysticetes are known to emit sounds of low and even very low frequencies, down to the famous 20-Hz-signals. Because the mammalian ear is generally rather insensitive to low frequencies, the problem arises how these animals manage to hear them. Several mechanisms have been proposed, including the detection of resonances of the lungs. It is the purpose of this presentation, however, to show that the cetacean ear is structurally specialized for the detection of low frequencies.

Before dealing with the hearing organ itself, it is necessary to point out that mysticetes have structurally modified their entire skull in order to suppress resonances. This is mainly achieved by soft connections between the long and slender bony elements. The long rostrum is only loosely tied to the occipital part of the skull, but this is only the most prominent modification.

The entire hearing organ is not fused to the skull, but decoupled, which is a further step towards suppressing the influence of vibrations of the skull on the hearing organ. In order to stabilize the decoupled hearing organ, a lot of dense bony material has been added, much more than necessary to house the inner ear and to support the middle ear. A schematic cross section is shown on the left of Fig. 1. Although modified, the same holds true for dolphins and porpoises as well.

A biomechanical analysis of the mysticete middle ear revealed that it contains a stapes complex much like the one in other mammals. The malleus-incus complex is basically the same as in dolphins with a Tursiops-type middle ear, indicating that it is a torsional system. Although no detailed measurements have been made it is obvious from the stiffness and the mass of both subsystems that they are tuned to higher frequencies: probably the natural frequency of the stapes complex is in the range of a few kHz, and that of the malleus-incus unit at several hundred Hz. Both components are therefore not well suited as receivers for the very low frequencies mentioned.

The cetacean ear differs from that in other mammals in that it is clearly divided into two major components which are both bulky and massive. As shown on the right of Fig. 1, one is the periotic element which houses the cochlea, while the other is the tympanic bone to which the malleus is fused (a situation quite normal for most mammals). In mysticetes the Tympanic is fused to the Periotic

891

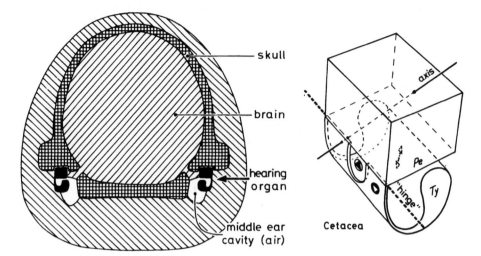

Fig. 1

left: schematic cross section through the cetacean head (lower
          jaw omitted) showing the isolation of the hearing organ.

right: schematic drawing illustrating the relation between the
          periotic bone (Pe) and the tympanic bone (Ty). The term
          "axis" refers to the rotational axis of the malleus-incus
          complex. Periotic and Tympanic can vibrate relative to
          each other around the "hinge" indicated. However, it is
          actually a flexional vibration.

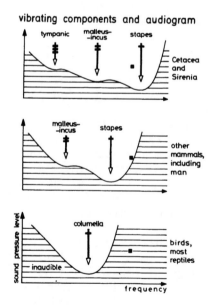

Fig. 2

simplified relation between the
vibrating elements of the middle
ear and the shape of the audiogram.
In man there are two vibrating
components: the malleus-incus
complex and the stapes. In
cetaceans the tympanic bone is
a third vibrating component. All
three subsystems combined increase
the frequency response of the ear.
(Sirenians have the same potential
but they do not seem to use it).

Fig. 1, right, and Fig. 2 are
reproduced from Fleischer (1978)
with the permission of Springer-
Verlag.

via two thin bony pedicles at the lateral side, while the medial
rim consists of a bulky mass of extremely dense bone. Because of
this configuration both elements, Periotic and Tympanic together,
vibrate relative to each other at low frequencies. The mode of
vibration can be visualized by the "hinge" indicated on the right
of Fig. 1, although it has to be emphasized that it is a flexional
vibration. (In dolphins and porpoises the situation is essentially
identical. A minor difference, which does not effect the functional
principle, is that in many of them there is no rigid fusion between
Periotic and Tympanic. Instead, this connection consists of a soft
coupling, via specialized contact areas and an intercalated layer
of cartilage).

An arrangement as the one just described can be simplified by
assuming that the Periotic is practically immobile during the vibration of the Tympanic, due to the inertia of its mass. The natural
frequency of the Tympanic then depends on the flexional stiffness
of the connection between Periotic and Tympanic, as well as on the
mass and the distribution of the mass of the Tympanic. It is obvious
that the development of the bulky medial rim of the Tympanic is a
measure to tune the system to low frequencies by shifting the center
of mass away from the "hinge", and by increasing the amount of mass.
The effect on the frequency response is the same as mass-loading a
tuning fork. (In odontocetes the flexional stiffness of the connection between Periotic and Tympanic is smallet, but the mass is also
smaller. However, basically the configuration is the same, making
it also a receiver for low frequencies, although it is very likely
not tuned to such extremely low frequencies as in mysticetes.

Preliminary experiments with a preparation of the hearing organ
from Balaenoptera physalus (fin whale) showed that the natural frequency of the Tympanic, relative to the Periotic, is at 29 Hz, thus
making it an effective receiver for very low frequencies. Because
the malleus is fused to the Tympanic vibrations of the latter will
be forced upon the malleus-incus unit, which in turn drives the
stapes as usual. Hence, vibrations of the Tympanic, relative to
the Periotic, will be transmitted to the cochlea via the ossicular
chain. The net effect of such a structural specialization is an
expansion of the frequency range of hearing, as shown in Fig. 2.
It enables the mysticetes to achieve a great sensitivity for very
low frequencies and it permits odontocetes to tune their ossicular
chain to high frequencies, while still retaining sensitivity for
low frequencies.

Fleischer, G., 1976, Uber die Beziehungen zwischen Hoervermoegen
     und Schaedelbau bei Walen. Saeugetierkundl. Mitteilungen
     (Munich), 24, 48-59.

Fleischer, G., 1978, Evolutionary Principles of the Mammian
     Middle Ear. Advances in Anatomy, Embrol., and Cell Biology,
     (Springer-Verlag; Berlin, Heidelberg, New York) 55:5.

MORPHOLOGICAL ADAPTATIONS OF THE SOUND CONDUCTING APPARATUS IN ECHOLOCATING MAMMALS

Gerald Fleischer

Echolocating mammals, bats, porpoises and dolphins, have a sound conducting apparatus in their middle ear which is fundamentally similar. In both groups there is a tympanic membrane, or its equivalent, an ossicular chain, composed of malleus, incus, and stapes, as well as two well-developed middle ear muscles. Also in both groups the malleus is fused to the tympanic bone, the auditory ossicles are separated from each other by thin layers of soft tissue, and the middle ear cavity is filled with air. In appearance, however, the sound conducting apparatus in bats is very much different from that in odontocetes.

An analysis of the sound conducting apparatus across mammals revealed that echolocators are only specialized as far as the tuning of the middle ear to the frequency range of hearing, and to a lesser extent the development of the middle ear muscles is concerned. Of critical importance is the size of the entire ear (not the size of the animal) and whether it operates in air or underwater. Minute middle ears, such as in bats, need special design to overcome problems in detecting low frequencies. The relatively large ears of the odontocetes, on the other hand, are engineered to overcome problems with the conduction of high frequencies. Furthermore, especially in cetaceans, the ear is decoupled from the skull as a measure towards vibrational isolation of the ears.

In the following section an outline of some aspects of the biomechanics of the sound conducting apparatus in echolocators is presented, as part of the evolutionary radiation of the mammalian middle ear.

The innermost element of the ossicular chain is the stapes complex which is a mass-spring system, tuned to the frequency range of the greatest sensitivity of the ear. Mass is the bony material of the stapes, while the spring is represented by the annular ligament which anchors the stapes within the oval window. In porpoises and dolphins the stapes appears to be immobile, because the spring (annular ligament) has to be so stiff, in order to give the stapes the natural frequency of 60 kHz, or more. In bats the anchoring of the stapes is not so extremely firm because its mass is much smaller. In summary, the natural frequency of the stapes is determined by the stiffness of the annular ligament, and by the mass-load (stapes) attached to it. It is a translatory system, i.e., it moves hence and forth like a piston.

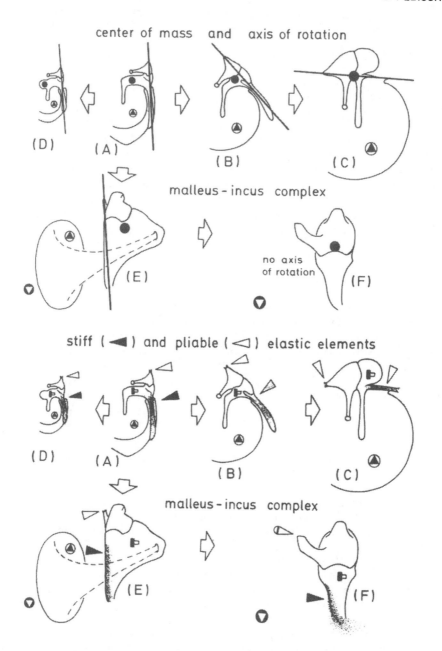

Fig. 1.   Evolutionary radiation of the malleus-incus complex in
          mammals.   (A) - ancestral type;   (B) - transitional type;
          (C) - freely mobile type;   (D) - microtype;   (E) - Tursiops-
          type;   (F) - Kogia-type.   Bats have the type (D), while
          dolphins have either type (E) or type (F).   The black
          triangle symbolizes the tympanic membrane.

influence of the orbicular apophysis

vibrational modes of the microtype

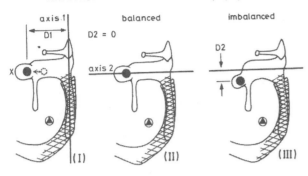

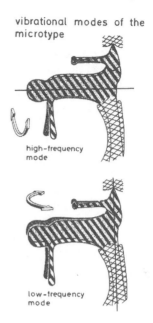

high-frequency mode

low-frequency mode

Fig. 2

Analysis of the malleus-incus complex in the microtype (bats). The orbicular apophysis (X) is a bulky mass of bone at the malleus. It determines the frequency response of the malleus-incus complex and its vibrational mode around two rotational axes.

(Fig. 1 and Fig. 2 are reproduced from Fleischer (1978) with the permission of Springer-Verlag; Berlin, Heidelberg, New York.)

Besides the stapes there is a second semi-independent system in the middle ear: the malleus-incus complex. It is a torsional element, the axis of which is shown at the top of Fig. 1. Its torsional stiffness is represented by the anchoring of the axis on both ends, and it varies greatly (bottom of Fig. 1). Even more important is the center of mass, relative to the axis of rotation, and it is this parameter which has the greatest influence on the tuning of the malleus-incus complex, according to the parallel-axis theorem. In bats (Fig.1 (D) and Fig. 2), this leads to the development of the orbicular apophysis, a bulky mass of bone at the base of the manubrium which determines both, the natural frequencies as well as the vibrational modes. Although different in appearance, odontocetes have a malleus-incus complex not unlike the one in bats (Fig. 1 (E) and (F)). In the Tursiops-type, a part of the tympanic membrane is stretched into the tough tympanic ligament which in turn rotates the malleus. In the Kogia-type, the tympanic membrane has been replaced by a ceramic-like plate of thin bone (white triangle). Without going into detail it has to be emphasized, however, that the ossicular chain in odontocetes is functional and that it operates according to the same principles as in bats or in man.

Fleischer, G., 1978, Evolutionary Principles of the Mammalian Middle Ear. Advances in Anatomy, Embrol., and Cell Biology, (Springer-Verlag; Berlin, Heidelberg, New York) 55:5.

ECHOLOCATION SIGNAL DESIGN AS A POTENTIAL COUNTER-COUNTERMEASURE

AGAINST MOTH AUDITION

James Howard Fullard and M. Brock Fenton

Much of the work on the role of audition in tympanate moths indicates that they possess ears which are tuned to the predominant frequencies in the echolocation signals of sympatric, insectivorous bats (Roeder, 1970). Since moths are more sensitive to certain frequencies there may be selection for bats whose echolocation signals tend to minimize acoustic reception by the tympanate moths they are hunting.

Electrophysiological studies of a variety of tympanate moths captured at the Station Biologique de Lamto in Côte d'Ivoire, West Africa were used to prepare audiograms for moth species representing seven families. Using the notodontid moth, Desmeocraera graminosa, we tested the responses of the auditory nerve to the echolocation signals of four sympatric, insectivorous bats: Pipistrellus nanus (Vespertilionidae), Hipposideros ruber (Hipposideridae), Nycteris macrotis (Nycteridae) and Rhinolophus landeri (Rhinolophidae) as they flew up to 3 m from the ear preparation. The echolocation signals of hunting bats in the area were monitored in the field with a broad-band ultrasonic microphone and portable frequency analyzer (Simmons et al., 1979).

The moths that we tested in Côte d'Ivoire were most sensitive to frequencies between 20 and 45 kHz, matching the frequencies most commonly monitored for echolocating bats in the field (Fig. 1). The ear preparations of D. graminosa displayed a differential ability to detect the cries of the four species of bats that flew in the room. P. nanus and R. landeri, species that employ high intensity (+ 90 dB re. 2 x 10^{-5} N/m^2, 10 cm) and, in the case of R. landeri, long duration (+ 25 msec) echolocation cries were both easily detected by the moth as they flew up to 3 m from the moth (maximum distance possible for the trial). H. ruber, a high frequency (+ 100 kHz), directional signal emitter was less conspic-uous to the moth and N. macrotis, which uses a low intensity (<50 dB) very short duration (2 msec) signal, was detected only when the bat was 10-15 cm from the moth and , as with H. ruber, only when flying directly towards the moth.

These data provide evidence that certain species of insect-ivorous bats may reduce the distance at which tympanate moths first detect them by emitting signals of a frequency and/or intensity that is difficult for the moth to hear. Coupled with the greater

899

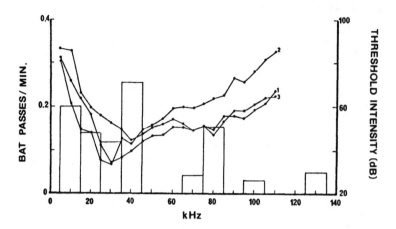

Fig. 1.   The relationship between the echolocation frequencies of
          hunting bats (bar graph) monitored in the field and the
          neurological audiograms (•—•) of sympatric moths. The
          moths tested are, 1 - Desmeocraera graminosa (n=10),
          2 - Miantochora interrupta (n=5), 3 - Rhodogastria vitrea
          (n=4).

flight speed of bats and a high degree of in-flight maneuverability
this reduced detection distance may provide some bats with a considerable advantage in catching tympanate moths.

        This research was supported by National Research Council of
Canada operating and equipment grants to MBF, and by a National
Research Council of Canada postgraduate scholarship to JHF.

REFERENCES

Roeder, K. D., 1970, Episodes in insect brains, Amer. Sci., 58:378.

Simmons, J. A., Fenton, M. B., Ferguson, W. R., Jutting, M., and
        Palin, J., 1979, Apparatus for research on animal ultra-
        sonic signals, Life Sci. Misc. Pub., R. Ontario Mus.
        (in press).

VOCALIZATIONS OF MALAYSIAN BATS (MICROCHIROPTERA AND MEGACHIROPTERA)

Edwin Gould

Repetitive, graded vocalizations occur in young bats during maternal-infant communication. In shrews, rodents, tenrecs, polar bears, ungulates, civet cats and some primates similar graded repetitive sounds are emitted and coupled to excitation levels. Occurrence of this coupling in such disparate groups suggests that phylogenetically the repetitive vocal emission system has been pervasive in placental mammals and that bats probably locked into an existing graded communication system for the development of echolocation.

During a visit to Malaysia I collected 10 species of bats that were either pregnant or nursing. Vocalizations of mothers and infants were recorded alone and during reunion. In general, calling rates of newborn infant bats are low; the rate increases from the first day (median=4 calls per sec; n=9 bats of 6 species) through the second week (median=6 calls per sec; n=11 bats of 6 species) and gradually wanes in the third week and near weaning (2 calls per sec; n=13 bats of 5 species) (Gould in press, Amer. Zool.).

In contrast to microchiropterans, Eonycteris spelaea (Megachiroptera) each night carries its infant out of the cave until the young infant is capable of flight. I netted nursing Eonycteris with infants attached to the nipple during their departure and return to the main entrance of Batu Cave. Infants detached from the mother at the cave were hand-held in front of the microphone and recorded with a Uher 4000 tape recorder. Sounds recorded later in the laboratory were indistinguishable from those recorded in the cave. Calling rates of Eonycteris infants were consistantly low compared to those of micro-chiropterans. Grinnell & Hagiwara (1972) reported that the temporal resolution and responsiveness to the second of a pair of sounds is the major receptive feature of the auditory system that is lacking in the non-echolocating bats. The inability of Eonycteris to emit calls of higher rate than about 2 per second may be another constraint that imposed limits on the ability of megachiropterans to echolocate with laryngeal sounds.

I recorded the vocalizations of mothers and infants of five Hipposideros species (Figure 1). Most infant Hipposideros sonar emissions emphasized harmonics that were lower than those of adult CF calls. As the infant matures, the second harmonic

is emphasized more and more; sounds higher than the second
harmonic of the typical CF call also occur. Infant non-sonar
calls emitted during reunion and isolation and in the absence
of any exploratory movements of the infants head or ears differ
from CF calls. Non-sonar calls of species such as H. armiger
and H. diadema differ in that H. diadema has a descending FM
sweep while H. armiger emits an ascending FM sweep (Figure 1).
I found nursing females of both species roosting in the same
cave. Likewise, about 4,000 nursing H. bicolor and more than
100 nursing H. cineraceus simultaneously occupied the same
roosting site; note their contrasting non-sonar calls.
Differences in neonatal calls presumably obviate confusion
during mother-infant reunion.

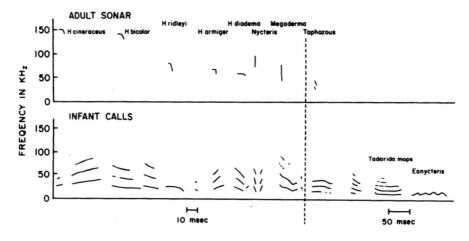

Figure 1. Sonar and non-sonar calls of 10 species of Malaysian
bats compared. No sonar calls were recorded for T. mops.
Eonycteris is a non-locater.

I recorded the adult sonar calls of five species of Hipposideros:
H. ridleyi, H. armiger, H. bicolor, H. cineraceus and H. diadema;
the first four had not been recorded previously. I plotted the
frequency of the CF portion of 16 species against forearm length
to determine whether vocal frequency and body size are related.
There is no linear relationship among the small species; it seems

that geographic distribution rather than size accounts for the grouping of small species with similar frequencies. The larger species that have rather widespread distribution form a second group of low frequency emitters. Variable frequency of the CF component occurs within and between populations in H. caffer and H. commersoni (See p. 125, Novick, 1978). Several members of this conference indicated their observation of variation in other species occurring in Figure 2. To what degree this tentative representation characterizes CF calls will depend on subsequent reports for other localities and populations.

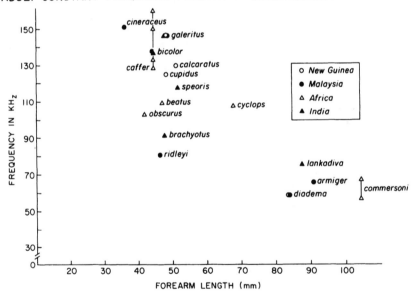

ADULT CONSTANT FREQUENCY PORTION OF HIPPOSIDEROS SONAR CALLS

Figure 2. The CF portion of sixteen species of Hipposideros are compared. Forearm length is a good measure of body size. For details, see 1977.

Acknowledgements: NSF Grant BMS 7202337A01; NIH Development Award; Fulbright Hays Program.

REFERENCES

Novick, A. 1977. Acoustic orientation, in: "Biology of bats", V.3, W. A. Wimsatt, ed., Academic Press, New York.

Grinnell, A. D. and S. Hagiwara 1972. Adaptations of the auditory nervous system for echolocation. Studies of New Guinea bats. Z. vergl. Physiol. 76, 41-81.

A RECONSTRUCTING TECHNIQUE FOR THE

NASAL AIR SACS SYSTEM IN TOOTHED WHALES

Vladimir S. Gurevich

Durable models of the nasal air sacs system of small delphinids were designed for three dimensional comparative anatomical study, and precise measurements and demonstrations were produced using a new technique. The few previous attempts to model the system were not completely successful (Gurevich, 1969; Mead, 1972; Schenkkan and Purves, 1973; Dormer, 1974). Latex, wax corrosion, and silicone rubber were used for reproducing the nasal air sacs system. Because the solvent in the previously used casting materials causes marked shrinkage (up to 70%) as it vaporizes, modeling of the small tubular and accessory sacs is unsatisfactory; the larger vestibular and premaxillary sacs may also be warped. Wax corrosion casts are very fragile, vinyl corrosions shrivel, and fusible alloy models are distorted.

In 1976, we began testing new materials. We were able to reproduce the entire nasal air sacs system and associated structures of Delphinus delphis, Stenella longirostris, and Tursiops gilli with Hasting's two-part epoxy resin, but could not obtain fully satisfactory results with fine details of muscle fibers, diagonal membranes, tubular and accessory sacs. Eventually, we discovered Batson's N 17 Anatomical Corrosion Compound, a product developed specifically for casting biological materials.

Batson's solution has many desirable characteristics for producing exact colored or colorless casts of any anatomical organ with an opening into which liquid plastic can be injected. The solution, a clear polyester resin, polymerizes at room temperature when mixed with one or more catalysts; shrinkage is about 15% and is easily injected by several means. Pigments are available if color is desired. Normal working time is about 20 minutes, but varies with amount of catalyst and color pigment used. The casts are relatively tough and easy to store and handle. The casting of the nasal air sacs system of nine species of delphinids Tursiops truncatus, T. gilli, Steno bredanensis, Delphinus delphis, Orcinus orca, Globicephala macrorhyncha, Phocoena phocoena, Phocoenoides dalli, Grampus griseus, and Kogia breviceps were produced with Batson's compound. Excellent details can be seen in the entire system including even such fine structures as the diagonal membrane, tubular and accessory sacs as well as their diverticuli and traces of the nasal plugs. Directions of the muscle fibres and their attachments also can be seen. The system of Kogia differs sufficiently from typical delphinids.

905

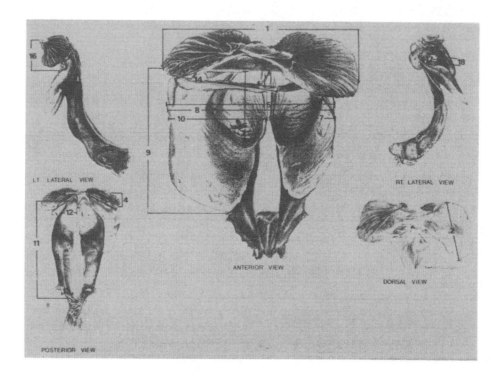

Figure 1

1. Width between external margins of vestibular sacs.
2. Length of vestibular sac.
3. Width of vestibular sac.
4. Height of vestibular sac.
5. Maximum width between external margins of nasal plugs.
6. Length of nasal plug.
7. Width of nasal plug.
8. Maximum width between external margins of premaxillary sacs.
9. Length of premaxillary sac.
10. Width of premaxillary sac.
11. Length of bony nàres.
12. Width of bony nares (top).
13. Width of bony nares (bottom).
14. Length of anterior horn of tubular sac.
15. Length of posterior horn of tubular sac.
16. Height of air sacs system.
17. Length of accessory sac.
18. Width of accessory sac.

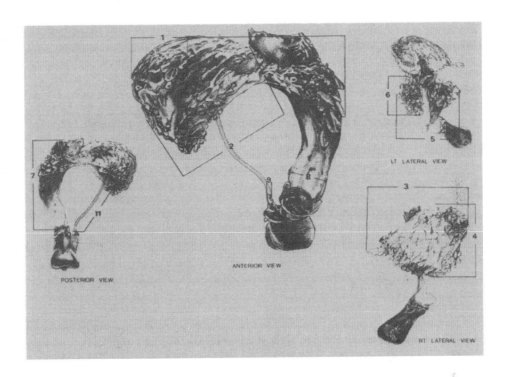

Figure 2

1. Width between external margins of vestibular sacs.
2. Length of vestibular sac.
3. Width of vestibular sac.
4. Height of vestibular sac.
5. Length of the left vestibular sac.
6. Width of left vestibular sac.
7. Length of left nasal bony passage.
8. Diameter of left nasal bony passage.
9. Diameter of right nasal bony passage.
10. Diameter of bony nares at bifurcation.
11. Projection of length of right nasal bony passage.

These anatomical structures are probably responsible for the sound producing in Odontoceti (Cetacea). Figures 1 and 2 show the casting models of the bottlenose dolphin (Tursiops truncatus) and pygmy sperm whale (Kogia breviceps) respectively. Numbers designate the area where the measurements were taken. The models, for the first time, allow us to conduct measurements of the nasal air sacs and associated structures which can be used for the future modeling of the dolphin's echolocator.

REFERENCES

Gurevich, V. S., 1969, Reconstruction technique of upper respiratory tract (nasal air sacs system) in Black Sea dolphins (unpublished).

Mead, J. G., 1972, Anatomy of the external nasal passages and facial complex in the Delphinidae (Mammalia: Cetacea). PH. D. Diss. University of Chicago.

Schenkkan, E. J., Purves, P. E., 1973, The comparative anatomy of the nasal tract and function of the spermaceti organ in the Physiteridae (Mammalia: Odontoceti), Bijdragen tot de Dierkunde, 43(1):93.

Dormer, K. J., 1974, Reconstruction technique for multiple castings of the sacs in silicone rubber. (Personal communication).